New Security Challenges Series

General Editor: **Stuart Croft**, Professor of International Securit... Politics and International Studies at the University of Wa... ESRC's New Security Challenges Programme.

The last decade demonstrated that threats to security vary ... manifestations, and that they invite interest and demand ... sciences, civil society and a very broad policy community. ... war was the primary objective, but with the end of the Col... defence as the centrepiece of international security agenda ... been, therefore, a significant shift in emphasis away from tra... security to a new agenda that talks of the softer side of securi... of human security, economic security and environmental security. The topical *New Security Challenges Series* reflects this pressing political and research agenda.

Titles include:

Abdul Haqq Baker
EXTREMISTS IN OUR MIDST
Confronting Terror

Robin Cameron
SUBJECTS OF SECURITY
Domestic Effects of Foreign Policy in the War on Terror

Sharyl Cross, Savo Kentera, R. Craig Nation and Radovan Vukadinovic (*editors*)
SHAPING SOUTH EAST EUROPE'S SECURITY COMMUNITY FOR THE TWENTY-FIRST CENTURY
Trust, Partnership, Integration

Sanjay Chaturvedi and Timothy Doyle
CLIMATE TERROR
A Critical Geopolitics of Climate Change

Tom Dyson and Theodore Konstadinides
EUROPEAN DEFENCE COOPERATION IN EU LAW AND IR THEORY

Håkan Edström, Janne Haaland Matlary and Magnus Petersson (*editors*)
NATO: THE POWER OF PARTNERSHIPS

Håkan Edström and Dennis Gyllensporre
POLITICAL ASPIRATIONS AND PERILS OF SECURITY
Unpacking the Military Strategy of the United Nations

Hakan Edström and Dennis Gyllensporre (*editors*)
PURSUING STRATEGY
NATO Operations from the Gulf War to Gaddafi

Hamed El-Said
NEW APPROACHES TO COUNTERING TERRORISM
Designing and Evaluating Counter Radicalization and De-Radicalization Programs

Philip Everts and Pierangelo Isernia
PUBLIC OPINION, TRANSATLANTIC RELATIONS AND THE USE OF FORCE

Adrian Gallagher
GENOCIDE AND ITS THREAT TO CONTEMPORARY INTERNATIONAL ORDER

Kevin Gillan, Jenny Pickerill and Frank Webster
ANTI-WAR ACTIVISM
New Media and Protest in the Information Age

James Gow and Ivan Zverzhanovski
SECURITY, DEMOCRACY AND WAR CRIMES
Security Sector Transformation in Serbia

Toni Haastrup
CHARTING TRANSFORMATION THROUGH SECURITY
Contemporary EU-Africa Relations

Ellen Hallams, Luca Ratti and Ben Zyla (*editors*)
NATO BEYOND 9/11
The Transformation of the Atlantic Alliance

Carolin Hilpert
STRATEGIC CULTURAL CHANGE AND THE CHALLENGE FOR SECURITY POLICY
Germany and the Bundeswehr's Deployment to Afghanistan

Christopher Hobbs, Matthew Moran and Daniel Salisbury (*editors*)
OPEN SOURCE INTELLIGENCE IN THE TWENTY-FIRST CENTURY
New Approaches and Opportunities

Paul Jackson and Peter Albrecht
RECONSTRUCTION SECURITY AFTER CONFLICT
Security Sector Reform in Sierra Leone

Janne Haaland Matlary
EUROPEAN UNION SECURITY DYNAMICS
In the New National Interest

Sebastian Mayer (*editor*)
NATO's POST-COLD WAR POLITICS
The Changing Provision of Security

Kevork Oskanian
FEAR, WEAKNESS AND POWER IN THE POST-SOVIET SOUTH CAUCASUS
A Theoretical and Empirical Analysis

Michael Pugh, Neil Cooper and Mandy Turner (*editors*)
WHOSE PEACE? CRITICAL PERSPECTIVES ON THE POLITICAL ECONOMY OF PEACEBUILDING

Nathan Roger
IMAGE WARFARE IN THE WAR ON TERROR

Aglaya Snetkov and Stephen Aris
THE REGIONAL DIMENSIONS TO SECURITY
Other Sides of Afghanistan

Holger Stritzel
SECURITY IN TRANSLATION
Securitization Theory and the Localization of Threat

Ali Tekin and Paul Andrew Williams
GEO-POLITICS OF THE EURO-ASIA ENERGY NEXUS
The European Union, Russia and Turkey

Natasha Underhill
COUNTERING GLOBAL TERRORISM AND INSURGENCY
Calculating the Risk of State-Failure in Afghanistan, Pakistan and Iraq

Aiden Warren and Ingvild Bode
GOVERNING THE USE-OF-FORCE IN INTERNATIONAL RELATIONS
The Post 9/11 Challenge on International Law

New Security Challenges Series
Series Standing Order ISBN 978–0–230–00216–6 (hardback) and
ISBN 978–0–230–00217–3 (paperback)

You can receive future titles in this series as they are published by placing a standing order. Please contact your bookseller or, in case of difficulty, write to us at the address below with your name and address, the title of the series and the ISBNs quoted above.

Customer Services Department, Macmillan Distribution Ltd, Houndmills, Basingstoke, Hampshire RG21 6XS, England

---

Climate Terror

A Critical Geopolitics of Climate Change

Sanjay Chaturvedi
Professor of Political Science, Centre for the Study of Geopolitics, Department of Political Science, Panjab University, India

and

Timothy Doyle
Professor of Politics and International Studies at the University of Adelaide; Distinguished Research Fellow of Indian Ocean Futures at Curtin University; and Chair of Politics and International Relations at Keele University

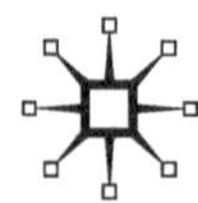

Softcover reprint of the hardcover 1st edition 2015 978-0-230-24961-5

First published 2015 by
PALGRAVE MACMILLAN

Palgrave Macmillan in the UK is an imprint of Macmillan Publishers Limited, registered in England, company number 785998, of Houndmills, Basingstoke, Hampshire RG21 6XS.

Palgrave Macmillan in the US is a division of St Martin's Press LLC, 175 Fifth Avenue, New York, NY 10010.

Palgrave Macmillan is the global academic imprint of the above companies and has companies and representatives throughout the world.

Palgrave® and Macmillan® are registered trademarks in the United States, the United Kingdom, Europe and other countries.

ISBN 978-0-230-24962-2 ISBN 978-1-137-31895-4 (eBook)
DOI 10.1057/9781137318954

This book is printed on paper suitable for recycling and made from fully managed and sustained forest sources. Logging, pulping and manufacturing processes are expected to conform to the environmental regulations of the country of origin.

A catalogue record for this book is available from the British Library.

Library of Congress Cataloging-in-Publication Data
Doyle, Timothy, 1960–
Climate terror : a critical geopolitics of climate change / Timothy Doyle, Sanjay Chaturvedi.
pages cm

1. Geopolitics—Developing countries. 2. Climatic changes—Developing countries. 3. Global warming—Developing countries. 4. Environmental justice—Developing countries. 5. Developing countries—Environmental conditions. 6. World politics—21st century I. Chaturvedi, Sanjay. II. Title.
QC903.2.D44D68 2015
363.738'74091724—dc23 2015001288

Typeset by MPS Limited, Chennai, India.

In fond memory of Sanjay's younger brother,
Rahul Chaturvedi

Contents

Preface

In early June 2013, a report for the World Bank by the Potsdam Institute for Climate Impact Research and Climate Analytics, entitled '*Turn Down the Heat: Climate Extremes, Regional Impacts, and the Case for Resilience*' (World Bank Report 2013) was published against the backdrop of extreme monsoons causing havoc in various parts of the Indian Himalayas. The report focuses on what it described as the three 'critical' regions of the world, 'sub-Saharan Africa, South-East Asia and South Asia'. It examines how impacts on agriculture production, water resources, coastal zone fisheries, and coastal safety are likely to multiply as global warming rises from its present level of 0.8°C up to 1.5°C, 2°C and 4°C, cautioning that as 'temperatures continue to rise, there is an increased risk of critical thresholds being breached. At such "tipping points", elements of human or natural systems – such as crop yields, coral reefs, and savanna grasslands – are pushed beyond critical thresholds, leading to abrupt system changes with negative effects on the goods and services they provide' (ibid.: xxiv).

On 15 June 2013, the Indian Meteorological Department reported the 'good' news of a 'plentiful' monsoon to the nation through extensive media reporting. Within a few days, 'Nature's fury' was unleashed in northern India, especially in the Himalayan states of Uttarakhand and Himachal Pradesh, turning the hope of a prosperous year into fear of large scale death and destruction, and also affecting thousands of pilgrims. The usual debate ensued between the environmentalists and government agencies over the precise nature of the calamity, with the former calling it 'manmade' and the latter describing it as 'natural'. As heavy monsoon rains destroyed lives (official figures being close to ten thousand), as well as infrastructure and local livelihoods, the longstanding mismatch between a high degree of vulnerability/risk and low levels of capacity/preparedness in one of the most disaster-prone parts of the globe was once again graphically, as well as painfully, exposed. It is worth noting that the above-cited World Bank report, while acknowledging that 'large uncertainty remains about the behavior of the Indian summer monsoon under global warming' (ibid.: 108), had cautioned that 'An abrupt change in the monsoon, for example, toward a drier, lower rainfall state, could precipitate a major crisis in South Asia, as evidenced by the anomalous monsoon of 2002, which caused the most serious drought in recent times – with rainfall about 209 percent below the long-term normal and food grain production reductions of about 10–15 percent compared to the average of the preceding decade' (ibid.).

The leading experts on 'natural' disasters in India were at pains to point out that three years ago an environmental assessment report by the Comptroller and Auditor General of India (CAG) had warned of the serious consequences of deforestation and fast multiplying hydropower projects. These were related to flash floods which could result in heavy losses of human lives and property. The report had noted that there were as many as 42 hydropower projects on Bhaghirathi and Alaknanda rivers, with 203 under various stages of construction and sanction, amounting to almost one power project every 5–7 km of the rivers.

The co-authors of this book come from both the 'Minority World' and the 'Majority World', joined together in the hope that this modest attempt will contribute to the mission of provoking a critical intervention by social sciences in the debate on climate change, which, despite growing evidence in favor of early and urgent action, is getting messier and murkier each passing day. At a time when the climate change debate is getting increasingly polarized between the accepters and deniers (despite their steadily declining number), social scientists cannot afford to take a backbench. We are also reminded that on the occasion of the award of the 2007 Nobel Peace Prize, the IPCC Chairman had expressed his disappointment over the relative silence of social sciences on climate change.

Let us begin with all emphasis at our command as social scientists that we do strongly acknowledge the force of the climate earth science and evidence furnished by it thus far, primarily through the IPCC reports, in favor of anthropogenic global warming. We also duly accept that global warming will cause wide-ranging, complex and compelling implications for humanity and its highly differentiated masses situated at various stages of uneven development (both material and human), and subjected to diverse political cultures and regimes of governance.

The evidence of global warming is most graphic and compelling in the case of all the Three Poles: the Arctic (French and Scott 2009; Chaturvedi 2012a), the Himalayas and Antarctica (Chaturvedi 2012b) but certainly not exhausted by them. We have burgeoning literature and fast proliferating print and visual media narratives on climate change trying to 'communicate' (though without much success so far) a sense of urgency, bordering on emergency, about the 'dangerous', 'disastrous' and 'catastrophic' consequences of climate change. We are also told by diverse actors, agencies, think tanks and media outlets dealing with the climate change issues that the worst victims of climate change will be the poor and the marginalized; a prospect worth noting in its own right but hardly a 'discovery' for those who have followed the exclusivist political, economic and cultural geographies of uneven 'development' unleashed by globalization and its more recent neoliberal avatars.

But, for a number of reasons, some of which are stated below, we tend to approach and analyze 'climate change' as comprising, but not confined to, global warming. In other words, we uphold that 'climate change', far from being a moment of rupture or radical departure, is a continuum marked by an ever shifting triad of statecraft and its political economies, nature and power. Climate change is not simply a matter of an abrupt, unprecedented 'global' manifestation of anthropogenic assault on nature in an abstract sense, with undifferentiated geographies of responsibility and accountability. In our view, it is better approached and analyzed as a messy convergence of various strands, paradoxes and dilemmas that have emanated from the reckless economic growth undertaken by the 'minority world' of the affluent and the influential. This has occurred in the context of what Lewis Mumford (1934) so aptly described as 'carboniferous capitalism', feeding into fossil-fuel driven urban civilizations resulting eventually in the 'end of nature', on the one hand, and rise of a 'global risk society' (Beck 2009), on the other.

It is equally important to bear in mind in our view – particularly while debating the 'mitigation' and 'adaptation' strategies for much 'feared' impacts and implications of climate change for different parts of the globe – that climate change is not alone on the post-cold war register of de-territorialized threats and dangers. It has joined an ever-expanding family of 'planetary' threats with allegedly trans-boundary spillover effects, said to be in dire need of a 'global' response and regulation, namely terrorism, epidemics, proliferation of weapons of mass destruction, the drugs trade, slavery, illegal migrations, etc.

There is no doubt that climate has been changing. But climate is not the only thing that is changing, or changing, for that matter, in complete isolation. To acknowledge this is to underline the importance of critical social science research on climate change. If climate change is about change in the human–nature interface then it is important to acknowledge that the history of the destruction and disappearance of nature in pursuit of primacy and domination, including the colonial chapter, is much longer than the history of global warming. The histories as well as the geographies of the domination of nature are still unfolding, with serious implications for human-livelihood security, and they must not be marginalized or erased by increasingly alarmist narratives of climate change.

As shown by Judith Shapiro (2001: 7) Mao's 'war on nature' during the Cultural Revolution was not the beginning, but a chapter in the longstanding history of 'exhausting the earth': 'a powerful national drive toward expansion, mastery, and resource exploitation, fueled by population growth and new technologies.' The point we are making in this book is somewhat paradoxical. No doubt the sooner the implications of global warming are acknowledged with a sense of urgency, the better for all, especially for the millions on the margins of affluence. But it is neither helpful nor desirable to downplay the fact that, as Bill Mckibben (2006: 148) points out, 'We live at a radical, unrealistic moment. We live at the end of nature ... when the most basic elements of our lives are changing.' Will 'climate' entirely displace and replace 'nature'? One of the reasons why climate change deserves immediate attention relates to another downplayed fact that climate change is now being seen as the ultimate symbol of a 'green identity' even (rather especially) in those countries where ecological irrationalities and injustices are deeply entrenched in the dominant practices of the political economy of statecraft and governance.

The word 'climate terror' is gaining increasing salience in the official and media pronouncements and can also be found on the websites of some of the think tanks. In some cases it has not only replaced the word 'fear' but also re-introduced the metaphors such as Mutually Assured Destruction (MAD) and Weapons of Mass Destruction (WMD). We wonder why. Some analysts would argue that geographical politics of fear, despite its seductive appeal in the face of highly complex geographies of transitions and transformations, does not seem to be working both for the 'Right' and the 'Left'. If so, then, we need to explore, in this case, the nature of this 'failure', and identify reasons behind the decline in the rhetorical utility of metaphors such as catastrophe and apocalypse. Or could it be that despite growing lamentation over the 'business as usual' attitude and approach to climate change, what we are witnessing is a rise of a new coalition geopolitics around a nascent but influential 'climate terror industry'. While lending most vociferous support for climate change mitigation and adaptation in order to save both the body and soul of the future citizen, this industry is not shy of seeking new business opportunities in the 'day after tomorrow'. Is the deployment of terror vocabulary to address climate change accidental or a part of refurbished designs and technologies of control, regulation and domination in a neo-liberal, post-political globalized world marked by profound asymmetries in terms of economic growth and development?

We argue and illustrate in this study that climate terror (an ensemble of various geographies of fear framed, flagged and deployed by various actors and agencies – state and non-state – in pursuit of their respective interests and agendas) is fast turning, at the same time, into a largely conservative grand-strategy deployed by faltering sovereign states, at various stages of a neo-liberal embrace, to discipline and regulate various faultlines in statecraft.

In the context of the complex and dynamic political geography of climate change, the processes of deterritorialization and reterritorialization, operating in conjunction, do not so much question the system of sovereign spaces as they reproduce it. Climate as a geopolitical space, therefore, is constantly moving *in* and *out* of physical-material geography. The imaginative geographies of climate change are always in the making, and intermittently assume territorial or nonterritorial forms depending upon the strategic convenience of the actors and their agendas concerned. In this new rhetorical map of the earth, the undifferentiated mass of humanity is imaginatively framed as integral to 'global soul' and cast within the shadow of a global enemy – climate – which is said to affect all (with the poor and the marginalized as the worst victims) but can only be interpreted and understood by a scientific and economic elite.

Post-colonial, post-partition South Asia (one of the most 'disaster-prone' regions in the world) is no exception to the global trend toward increasing de/re-territorialization as well as securitization of climate 'spaces'. For example, one of the most alarmist ways in which climate change is folded into a discourse of fear in support of various domestic and foreign policy agendas within Bangladesh and its neighbor India, is by referencing the 'problem' of *millions* of 'climate migrants' and 'climate refugees' – a 'problem' which then demands a geopolitical response.

The geopolitics of climate change in the foreseeable future will continue to oscillate between various imaginative geographies of fear and counter-imaginative geographies of hope, depending upon their ideological moorings and power-political agendas. It is worthwhile to explore the prospects for the latter and the role they could possibly play in approaching the issue of climate induced displacements from the angle of human security and human rights of the socially disadvantaged, dispossessed and displaced in this part of global South. This in turn would demand a relentless interrogation of the complex geography behind the politics of fear and the ways in which various parts of the globe, especially in global South, are *implicated in* and at the same time *induced by* doomy Malthusian scenarios of climate catastrophism.

Acknowledgments

Each of us owes individual and combined debts of gratitude.

Sanjay would like to thank Arun Grover, Vice Chancellor, Panjab University, Chandigarh, for his support and inspiration throughout the sabbatical he availed for this book writing. He would also like to thank his colleagues in the Department of Political Science for their encouragement and his research scholars for their valuable research assistance. A very special thanks to India China Institute (ICI), The New School New York (where Sanjay was a Third Cohort Fellow during for 2010–2012), especially its Director, Ashok Gurung, for extending full support to this book project. Particular thanks are extended to Jayanta Bandyopadhyay, Shikui Dong, Sze Ping Lo, Nidhi Srivinas and Victoria Marshal for sharing their knowledge and insights on issues related to environmental sustainability. Thanks also to Mahanirban Calcutta Research Group (CRG), especially Ranabir Samaddar, Paula Banerjee, Samir Das and Sabyasachi Basu Ray Chaudhury.

Tim wishes to express his gratitude and good wishes to colleagues and students of politics and international relations at the University of Adelaide in Australia, and at Keele University in the United Kingdom who contributed to, and took his courses on, Global Environmental Politics, Global South and International Political Economy. Particular thanks are accorded to Doug McEachern, Stoifan Darley, Jessica Hodgens, Brian Doherty, Sherilyn MacGregor, Adela Alfonsi, Phil Catney, John Vogler, Andy Dobson, Steve Quilley, Bulent Gokay, Andy Lockhart, Hayley Stevenson, Mohamed Salih and Barry Ryan. In more recent times, the support from Graham Seal of the Australia-Asia-Pacific Institute at Curtin University has been truly valued. Thanks also to Tim's colleagues in the Indo-Pacific Governance Research Centre at the University of Adelaide.

On a more personal note, Sanjay would also like to thank his wonderful family, especially his parents Shri Dinesh Chaturvedi, Smt. Indrani Chaturvedi and Shri Priyavrat Sharma for their blessings and guidance, wife Veena for her unconditional love and care, most adorable children Pranshu, Sukriti, Pranjal and Sameeha for their emotional support, and brotherly friend Manbir Singh for his invaluable affections. Tim would like to equally extend his love and warm appreciation to his parents,

Maureen and Seamus, and to his wife Fiona, and children Georgia, Matilda and Thomas.

Some ideas already expressed in print need to be acknowledged here.

Sanjay and Tim together both would like to acknowledge their heartfelt gratitude to Marcus Power, the editor of the special issue of *Geopolitics* on 'Geopolitics and Development' (Routledge, 15(3) 2010) and to Dennis Rumley, founding Chief Editor of the *Journal of the Indian Ocean Region* (Routledge, Special Issue on Climate Change and the Indian Ocean Region, 6(2) 2010) for encouraging us to submit articles that draw upon a much larger ongoing project that has eventually culminated in the form of this book.

We would also like to extend our gratitude to Carol Johnson and Juanita Elias of Universities of Adelaide and Warwick respectively, for asking us to develop some early ideas on these issues as evidenced in 'Critical Geopolitics of Climate Change and Australia's "Re-Engagement" with Asia', special edition, *Australian Journal of Political Science*, Taylor and Francis, 2010. Similar thanks must be accorded to John S. Dryzek, Richard B. Norgaard and David Schlosberg, editors of *The Oxford Handbook of Climate Change and Society* (OUP, 2011): in asking us for a contribution on 'climate refugees', it got us thinking more clearly about the different discourses on climate securitization.

Much appreciation is accorded to Brian Doherty at Keele University – Tim's fellow author on another recent Palgrave Macmillan book, *Environmentalism, Resistance and Solidarity: The Politics of Friends of the Earth International* (2014) – for allowing Tim to reproduce sections of writing and research derived from this study. Gratitude to another of Tim's Keele co-authors, Phil Catney, for allowing him to use insights derived from the writing of their joint article, 'The Welfare of the Now and the Green (Post) Politics of the Future', *Critical Social Policy*, vol. 31(2), 174–193, 2011. Finally, great thanks to Mohamed Salih from University of Rotterdam, and Tim's co-author Andy Lockhart, from University of Sheffield, for allowing him to reproduce elements of our shared argument also presented within, 'Local campaigns against Shell or transnational campaigns against climate change? From the Niger Delta, Nigeria to Rossport, Ireland,' in Mohamed Salih (ed.) *Local Climate Change and Society*, 138–204 (Routledge, 2013).

We both owe a very special gratitude to our brother and most respected colleague Dennis Rumley for having inspired us in so many ways. His support, insights and encouragement have been one of our biggest assets throughout the entirety of this book project. The three of us were part of the original team forming the Indian Ocean Research Group (IORG), and this friendship continues to bear remarkable fruit.

For Adela Alfonsi of the University of Adelaide, our appreciation cannot sufficiently be expressed in the form of a simple acknowledgment. Her editorial support, administrative muscle, and intellectual rigor, provided at just the right times, have been instrumental to this project's completion.

To Georgia Lawrence-Doyle, thanks also for the outstanding work in this book's early research phase.

We are truly beholden to the Editor of Palgrave *New Security Challenges Series*, Stuart Croft, Warwick University, for his initial interest and support of this book-length research project. We are also grateful to our wonderful colleagues at Palgrave Macmillan, Ellie Davey-Corrigan, Commissioning Editor for International Relations, and Hannah Kašpar and Emily Russell for their thorough and timely editorial advice and assistance. Thanks also to Alexandra Webster who originally commissioned this book project for Palgrave Macmillan.

We would like to acknowledge that significant parts of this book draw upon the research of a much larger three-year project entitled 'Building an Indian Ocean Region' (DP120101166), which is funded by the Australian Research Council (ARC) Discovery Projects Scheme for funding 2012/2015.

1
An Introduction: A Critical Geopolitics of 'Climate Fear/Terror': Roots, Routes and Rhetoric

Introduction

On the 'doomsday clock' of the *Bulletin of the Atomic Scientists*, which intends to caution 'how close humanity is to catastrophic destruction', 'climate change' joins the other two alarmist categories, namely 'nuclear,' and 'biosecurity'. At the same time, there is a grudging acknowledgement of the fact, at least by some, that the geopolitics of fear, deployed at diverse sites by different agencies – individually and/or collectively – in pursuit of various interests and agendas, has failed to yield the desired results, including a change in public and private behavior and for that matter the ushering in of radical social movements (Lilley 2012). On the contrary, it appears to have resulted in 'catastrophe fatigue, the paralyzing effects of fear, the pairing of overwhelmingly bleak analysis *with inadequate solutions, and a misunderstanding of the process of politicization*' (ibid.:16; emphasis added). Could this be the reason that some of these multifaceted discourses of fear – that somehow remain open to political contestation and interrogation – are now being scaled up and upgraded by various regulatory agencies and alliances to the discourse of 'climate terror'? This discourse can only have counter-terror as its Other in order to completely erase the hope (the Other of fear) of re-ordering and regulating spaces and societies allegedly more vulnerable to climate change and its threat-multiplying effects. Is climate terror an apparatus of govern-mentality that aims at erasing not only the collective memories of historically perpetuated environmental injustices by the powers that be, but also hopes to contain growing resistance in various parts of the globe (especially the global North) against the emerging architecture of domination and dependencies? Of course, the separation between North and South is a useful category, marking out the affluent lives of the minority versus those of the less affluent majority. But, like any border (as mentioned in the preface of this book), it is drawn subjectively and imperfectly to demarcate territories (Doyle and Chaturvedi 2010).

As pointed out by Eddie Yuen (2012: 37), 'The prevalence of fear-based catastrophism reveals the depth of acceptance of the assumption of rational choice theory in both natural and social sciences. The assumption of a certain kind of instrumental rationality undergirds the delusional belief that if only people could understand the scientific facts, they will change their behavior and trust the experts.' This entails putting a heavy gloss over a set of deeply political and politicizing questions at a time when the lure of the post-political is gaining traction but not without inviting a micro geopolitics of resistance.

Can we say that 'climate terror' is the accumulated, collective outcome of steadily proliferating fears, with each fear serving to endow its anticipatory regimes with 'expertise' and 'clinical authority'? What kind of language, imaginaries and metaphors are being deployed to frame and communicate climate change, by whom and why? What is the politics behind the written geographies of climate change, and how and why are largely Afro-Asian places and people being framed and implicated in various *geographies* of catastrophe and fear? What are the implications that these discourses carry for understanding climate change and choosing 'appropriate' policy options and responses? What minimum ethical principles related to equity are needed to ensure that the impacts of policies to address climate change are perceived as equitable by key stakeholders (Giddens 2008: 4) and do not result in further marginalization of the much less fortunate losers of corporation globalization?

Structure of the book and its order of exposition

In Chapter 1 of the book, our key engagement focusing the rhetoric of 'climate terror' is pitched at a number of theoretical perspectives that inform 'critical geopolitics' essentially as a relentless interrogation of the politics and even depoliticizing politics of domination. While being 'critical', we do not dismiss state-centric classical geopolitics out of hand since we believe that nation-states, irrespective of their geo-economic and geopolitical locations, continue to matter a good deal in international geopolitical economy. For example, low-lying Bangladesh and Maldives are 'vulnerable', not only because of their vulnerable physical geographies, but also because of their geopolitical location in the global South.

In this chapter then, we provide a brief historical overview of the engagement of both classical and critical geopolitics with categories of 'geography', 'nature' and 'environment' and their interplay with statecraft. In addition, we argue for expanding the nature and scope of critical geopolitics through engagement with various 'critical' perspectives (in contrast with conventional wisdoms) of social sciences and humanities; as well as convergent critical perspectives around the notions of space, scale and power.

In Chapter 2, we investigate the anxieties, uncertainties and denials associated with both the construction and broadcasting of climate 'Science' and the politics of knowledge. In this vein, we research the relationship between climate science and the politics of fear, looking at the categories of control and the hegemony of the natural sciences; the science and power of climate change paradigms and mythologies; and diverse cultural understandings found in indigenous knowledge. In this vein, we briefly touch upon controversies over the melting of Himalayan glaciers and pre-Copenhagen East Anglia.

Additionally, we trace the changing discourses of early environmental movements which often directly challenged 'Enlightenment Science' and its myths of progress-at-all-costs, to more recent green movements which have become 'ecologically modernized', now advocating concepts such as sustainable development and win-win-win political games. The notions of ecological modernization, sustainable development and the scientific knowledge which inform them, we argue, stand co-opted and depoliticized, with the zero-sum games of the finite, hard-earthers now replaced with *Plasticine Anthropocene* understandings of the Earth as infinitely malleable.

One key argument informing this work is the concept of the post-political, and in Chapters 3 and 4, we discuss the relationships between the post-political and the discourses of neo-liberal globalization. In Chapter 3, after introducing the post-political at some length, we focus on *climate territories* with their marginalized geographies inherent in post-political constructions of geopolitics, leading us to a discussion of Climate and the Anthropocene (exploring the 'boundless' nature of climate change) and critically argue against what we contend is the construction of a *'Global Soul'* for 'Global South' by Northern elites. Finally, in this light, we study space, scale and the politics of making/unmaking places, and in the latter pages we utilize the case of India to illuminate our purposes.

Chapter 4 moves to Climate 'Markets', for it is within neo-liberal tropes about climate change that the concept of a post-political, post-material geopolitical reality is most usually found. In specific terms, we endeavor to expose the emergence of neoliberal green economics, providing examples and differences between ETS systems versus carbon taxes (the role of states versus market driven schemes). Of course, climate change has become so omniscient that other economic ideologies and approaches have also responded to its call (not just the neo-liberals), and we seek to provide examples of mercantilist and more classically liberal approaches to climate economics. We strongly argue, however, that climate change emerges from a conservative, right-wing morality, so its relationship with right-wing economic doctrine (in all its forms) is hardly surprising.

Chapters 5 and 6 move to the world of securities and securitization. In Chapter 5, we focus on *climate borders*: securitization, flows, migration and refugees. In this analysis we cast our critical gaze on the construction of the climate 'terrorist'; climate cores and peripheries; mobility and circulations; and investigate the realist and/or neo-Hobbesian literature on climate 'wars' and conflicts. For much of this chapter, however, our focus is largely on the *climate refugee*, and in this manner, we concentrate on displacements and migrations, using a detailed Bangladeshi case study to ground our theoretical musings.

Chapter 6 takes this discussion on security one step forward and reifies a theme which has developed right throughout the work: the close connection between geo-securities and geo-economics. Post September 11, and post the 2007–8 financial crisis, in a geopolitical sense, financial limitations continue to justify neo-liberal responses to global security. The Earth (or the Climate) is now increasingly seen by global elites as little more than a collective of post-political citizen/consumers of the core, whose interests to trade in marketplaces need more amorphous and less permanent forms of 'protection' (provided by nation-states in the past) from those dwelling in the *black holes* of market periphery. This chapter looks at the manner in which climate security has been militarized. Case studies of the United States military, and its 'green defense' projects are provided in a new and powerful geo-economic/geo-security region/non-region now referred to as the *Indo-Pacific*.

In the penultimate chapter, Chapter 7, we extend our thoughts beyond the domain of governments and corporations. We ask the question: how have social movements, non-government organizations, unions and churches responded to the climate change phenomenon? In particular, we provide some explanations as to how more 'emancipatory' groups and networks have responded to concepts such as climate justice and climate debt; how these groups who, in spite of also being co-opted by the might of climate change dogma, have attempted to use this global *climate moment* for more democratic purposes. Although we provide brief cases from both Church groups and the Union movement, we concentrate on how green movements themselves have been impacted upon, and for much of the chapter, we offer an analysis of the largest global green organization: Friends of the Earth International, with branches in over 70 countries, in the global North and South.

In Chapter 8, we conclude with our understandings of *climate futures*. We provide an overview of climate diplomacy, and investigate notions of common but differentiated responsibilities, respective capabilities and global governance. We also touch upon the geographical politics behind climate engineering.

We revisit notions of power, knowledge and technology and, in the end, advocate the resistance of artificial climate futures. This discussion leads us to one of our final questions: Can climate, as a set of discourses, be utilized for emancipatory ends or, ultimately, is the climate story, regardless of its diverse intentions, a discourse now captured by the affluent North to control the development of the global South? In short, has the emancipatory moment now passed or is there still hope for the re-emergence of subaltern perspectives on climate futures?

Toward a critical geopolitics of anthropocence, global warming and climate change

What emerged during the 1980s within the sub-discipline of political geography was a new approach called 'critical geopolitics' with the overall objective of liberating geographical knowledge(s) from the old and the new imperial geopolitics of domination. In the words of one of the leading proponents of critical approaches, 'The focus of critical geopolitics is on exposing the plays of power involved in grand geopolitical schemes' (Ó Tuathail 1992: 439). It is aimed at relentless interrogation of the 'power of certain national security elites to represent the nature and dilemmas of international politics in particular ways. These representational practices of national security intellectuals generate particular "scripts" of international politics concerning places, peoples and issues. Such scripts are part of the make-up by which hegemony is deployed in international systems' (ibid.: 438).

As pointed out by John Agnew (2010: 569), 'The hegemonic calculus of the past 200 years has involved the imposition of a set of normative rules and practical constraints on states and other actors, reflecting the uneven distribution of global power and a common "script" of world politics thereby written more in some places than in others. Though this script has had powerful continuities to its core themes, it has also involved important shifts over time with the rise and fall of dominant actors who have brought different conceptions and practices to bear within it.' Agnew points out that so far most of the scholarship in critical geopolitics has engaged with contemporary United States and the European colonial powers, 'often as if they were the sole active forces in world politics toying with the docile masses in the rest of the world' (ibid.).

We have been conscious of, and inspired by, the insights offered by one of the leading proponents of critical geopolitics, Simon Dalby. According to Dalby, geographical knowledges have been used and abused in the past for the purposes of so-called 'discovery', enclosures and expropriation during the colonial times and will continue to serve various imperial impulses and neo-colonial projects in various parts of the globe. Matthew Sparke's (2005, 2007) insistence on geographers using a 'post-foundational ethic as our guiding principle and collectively challenge the taken for grantedness of these practices', points out Dalby, could be used to question and critique 'the violence and transformations we have unloosed in the biosphere' (Dalby 2010: 280).

> This is especially important in the circumstances of *our* increasingly artificial existence in the urbanized world of the Anthropocene where we are collectively remaking our fate in ways that render traditional notions of a separate nature or an external environment untenable premises for discussing the earth as humanity's home ... Linking the spatial and natural themes in the discipline puts the most basic questions of politics at the heart of geographical considerations. Are *we* then to understand ourselves as on earth, squabbling over control of discrete territories and threatening massive violence to our putative rivals in other sovereign spaces, or are we to understand our fate as increasingly a matter of reorganizing a dynamic biosphere in which *we* all dwell? (ibid.; emphasis added)

From an ethical-normative standpoint, the authors of this book, with a 'political science' background, are inclined to be a part of *WE* that Dalby is alluding to as a critical geographer. Yet we feel slightly uncomfortable with a universalized notion of 'we' (while aspiring toward that state of collective socio-spatial consciousness) and would therefore like to introduce in this study critical geopolitical perspectives *on* and *from* the global South.

We do agree with Matt Sparke's contention that geographical grounds of fear and hope need to be critically examined, if only for the reason that these 'two are huge swirling compulsions with enormous implications for the lives and deaths of every living thing on the planet. False hopes and groundless fears can be of dreadful deadly consequences. And yet justified fears when combined with sensible hopes can open new possibilities and thereby help mobilize change for the better' (Sparke 2007: 338). He goes on to explain:

> We can usefully come to terms with the *double vision of fear and hope* through recourse to arguments about geopolitical scripting and geoeconomic 'enframing'. Critical investigations of the imaginative geographies produced by geopolitics and geoeconomics help us to understand how the fears and hopes of those who promoted the war were both groundless and yet at the same time ground changing. (ibid.: 339)

Sparke's insistence 'that geopolitics and geoeconomics are better understood as *geostrategic discourses*' (ibid.) appears to be quite relevant in the case of the climate change metanarrative. Various geopolitical and geoeconomic strands of the narrative are unfolding – and in some cases in a rather overlapping manner – at various sites, including: national defense-security establishments, ministries and departments dealing with earth sciences and environment; corporation-government partnerships engaged in carbon trading of increasingly territorialized carbon sinks (Lovbrand and Stripple 2006); religious groups; trade unions; nuclear as well as fossil-fuel industries; environmental NGOs with new climate change portfolios; and insurance companies, to name just a few.

Even though the issues raised by Sparke pertain to the Iraq war and not to the geoeconomic framings and geopolitical scripts of climate change, they are helpful in understanding the contradictory double vision of American discourses on climate change. The following official statement released on the eve of President Obama's pronouncement of national climate action policy through a short video addressed to the citizens of the United States resonates this double vision, anchored in rather multiple oscillating reasonings, quite graphically.

> I'll lay out my vision for where I believe we need to go – a national plan to reduce carbon pollution, prepare our country for the impacts

> of climate change, and lead global efforts to fight it. This is a serious challenge – but it's one uniquely suited to America's strength. We'll need scientists to design new fuels, and farmers to grow them. We'll need engineers to devise new sources of energy, and businesses to make and sell them. We'll need workers to build the foundation for a clean energy economy. And we'll need all of us, as citizens, to do our part to preserve God's creation for future generations – our forests and waterways, our croplands and snowcapped peaks. There's no single step that can reverse the effects of climate change. But when it comes to the world we leave our children, we owe it to them to do what we can. So I hope you'll share this message with your friends. Because this is a challenge that affects everyone – and we all have a stake in solving it together. (cited in Chris Good 2013)

In the deployment of geopolitics of fear, 'imaginative geographies' play an important role. According to Derek Gregory (2004: 17) 'Imaginative geographies imply, 'Representations of other places – of peoples and landscapes, cultures and 'natures' – that articulate the desires, fantasies and fears of their authors and the grids of power between them and their "Others"'. Gregory's critical engagement with a 'colonial present' shows how contemporary geopolitical discourses of fear and enmity have both roots and routes in imperialism (ibid.). It is equally useful to note that many imaginative geographies of imperial Orientalism, systematically critiqued by Edward Said, were refurbished and deployed in order to both stage and legitimize the 'War on Terror' (ibid.). With the help of global media or what some scholars have termed as the CNN effect (Gilboa 2005), 'the geopolitical scripts about despotic, hate-filled Orientals served to provide the fear-filled justification for treating whole communities as if they lay outside the bounds of humanity' (Sparke 2007: 343).

We will argue and illustrate in later sections of this book that the ascendance of climate terror discourse, and the ways in which imaginative geographies, linking the consequences of climate change to various facets of the war on terror, are being manufactured by various military-security think tanks. They reinforce the emphasis placed by Allan Pred (2007) on the performative aspect of the geopolitics of fear. Taking note of various opinion polls is no doubt helpful in some ways, but is not enough since it conceals 'how the enunciation of fear became a performance of sovereignty and governmentality at the same time but in different places' (Sparke 2007: 343).

The performative aspect of a geopolitics of fear needs further scrutiny in order to expose the violence (both epistemic and structural) that often accompanies various technologies of control, supported by a curious mix of persuasion and coercion, directed at remaking and reordering the real so that it fits into the imagined. Later in this book, in our analysis of the rhetoric and reality of climate change-induced displacements and migrations, we will show how the most 'vulnerable' in various parts of the world, especially the global South, are being discursively transformed into the most 'dangerous'.

Susan Roberts, Anna Secor and Matthew Sparke (2003: 886) draw attention to 'a more widespread form of *neoliberal geopolitics* implicated in the war-making.' What we find in this geopolitical world vision is the 'neoliberal idealism about the virtues of free markets, openness, and global economic integration' linked to 'an extreme form of American unilateralism' (ibid.). Furthermore, in contrast to the 'ideological geopolitics' of the Cold War era (Agnew 1998), where danger was perceived as 'something that should be contained at a disconnected distance', today it is the disconnection from the US lead globalization project that is understood to represent dangers of various kinds. In Chapter 6, we return to a detailed discussion of how a neoliberal geopolitical response appears to insist on *enforcing* reconnection, in the context of highly uneven geographies of globalization, through a hybrid strategy of persuasion and coercion (Sparke 2013).

Geopolitics of climate fear: sites and sights

The key concern here is with the fear-inducing narratives constructed and used by policy makers and politicians in pursuit of their so-called 'national interests' and related foreign policy-diplomatic agendas. These can be found in the speeches delivered and/or statements made by the politicians, including for example those posted as a matter of routine on the official websites of the ministries of 'foreign' or 'external' affairs. The key challenge here is to discern and deconstruct the 'practical geopolitical reasoning' in a foreign policy discourse (see Ó Tuathail and Agnew 1992); ' ... reasoning by means of consensual and unremarkable assumptions about places and their particular identities' (ibid.: 96). For example, through the 2002 State of the Union address, former US president George W. Bush could evoke the imaginative geographies of an 'axis of evil' by naming Iran, Iraq, and North Korea in conjunction with their alleged 'terrorist allies'.

The scenarios and spectacles outlined by practical geopolitics, often with the help of certain metaphors (e.g. 'rogue' states, 'domino effects'), are not always explicit or alarmist. The accumulative effect of repetitive utterances, however, need to be carefully mapped because 'the power of practical geopolitics is in its banality. Geopolitical ideas often appear so ordinary as to be invisible ... The repetition of geopolitical ideas within the practical performance of politics serves to naturalize certain categorizations of the world: for example, developed/less developed, core/periphery, or simply "us" and "them". These phrases may seem innocuous, but they are affirming particular political perspectives and legitimizing foreign policy decisions' (Painter and Jeffrey 2009: 208). In our analysis of the geopolitics of climate change we would be deploying critical geopolitical perspectives in order to expose what John Agnew describes as 'the techniques of concealment and spatial fixing associated with the dual geopolitical disciplining and intellectual naturalization of the world political map' (Agnew 2010: 570).

The manner in which masses are being socialized into dominant representations of other places and peoples (positive or negative) through media, cinema, cartoons, books and magazines is the subject matter of popular geopolitics. As the political geographer Joanne Sharp (2000: 31) puts it, 'hegemony is constructed not only through political ideologies but also, more immediately, through detailed scripting of some of the most ordinary and mundane aspects of everyday life.' Sharp (ibid.) has shown how, during the Cold War era, various contributions to the *Reader's Digest* highly exaggerated the 'red' threat from the Soviet Union, called by the former U.S. president Ronald Regan the 'evil empire'.

In our view, critical geopolitics needs to pay a far more serious and systematic attention to how imaginative geographies, anchored in fear, are deployed at the service of objectification, embodiment and instrumentalization of abstract risks, threats and dangers. The strategies deployed to objectify and embody abstractions through evidence deserve scrutiny. These imaginative, imagined, ideational, and emotional geographies in some cases (as demonstrated in Chapter 5) could be far more complex and compelling than the material geographies.

Critical geopolitical perspectives on mapping risks further reveal that discourses framing local and distant dangers in neo-Malthusian terms are often anchored in geographically vague and imprecise reformulations in spatial terms. In geopolitical terms, then, 'the environmental challenges of the 21st century represent a dialectic of territorialization and deterritorialization, a mixture of spatial fixity and unfixity. It is here, though, that things start to get really interesting and complicated' (Castree 2003: 427). Ecological degradation or climate change is a problem frequently attributed to the 'over-populated' global 'South'. What is very much in vogue here is the geopolitical impulse that divides the world into blocks with specific attributes. While modernity endangers itself in several ways, as brought out by Ulrich Beck (1992) in his theory of risk society, most of these dangers are still frequently framed and flagged in reference to 'external' causes. As pointed out by Simon Dalby (2003: 445), ' ... the geography of specifying the environmental threat as somehow external obscures the fact that these threats might better be understood in terms of the unintended long-term and long-distance consequences of our own action.'

A critical geopolitics of climate fear, in our view, should pay equal attention to both the geographical and political dimensions. It looks like the concept of fear has received much greater attention of geographers than of political scientists. Writing on the subject in the immediate aftermath of 9/11, John Keane (2001) expressed concern over the huge neglect of the concept of fear by political scientists, and raised a number of compelling issues with regard to fear and democracy, which deserve reflection in the context of climate change. Drawing attention to the 'fear economy' that followed, he argued that, 'the spread of fear outward from the United States, helped by the rapid circulation across borders of images, sounds and reported speech, arguably represents a new phase of globalization of fear that began after World War One and was reinforced by the events of the following world war and the invention and deployment of nuclear weapons' (Keane 2001: 3). While giving due credit to Montesquieu for drawing attention to how despotic regimes represent 'a type of arbitrary rule structured by fear' (ibid.: 4), Keane critically engages with the contention that established democracies tend to 'privatize' fear and are in a better position to reduce and control fears through innovative ways.

While agreeing with Montesquieu that different political systems have exhibited remarkably different forms and intensities of fear, Keane argues that the phenomenon of fear, as a 'particular type of physical and bodily abreaction of an individual and group', should be approached and analyzed 'within a triangle of inter-related experiences', namely 'objective circumstances', 'subjective systems' and 'abreactions'. What goes on within this triangle is a multitude of reactions triggered by wide-ranging circumstances sensed by individuals and groups as 'ill-boding, sinister, menacing, and perhaps even life-threatening' (ibid.: 18). Keane calls for a critical scrutiny of how several counter-trends – including institutions of statecraft flagging and fanning the fear of war, civil societies mourning 'the loss of imagined stable communities of the past', and media fascinated with fear consciously or unconsciously – ensure that fears, rather than being trivialized and eliminated, become a permanent 'public problem within both potential and actually existing democracies'. Even though located a considerable geographical distance from the theatres of 'just wars', the so-called tamed democratic zone of peace, prosperity and stability and its inhabitants find themselves in highly discomforting geopolitical proximity to the complex geometry of globalized fears centered on allegedly wild and unpredictable parts of the global South. The boundary between the public and the private domains of fears is therefore increasingly difficult to maintain.

> To the extent that fears once suffered in private have come to be perceived and dealt with as public problems, the ground is prepared for the understanding of fear as contingent, as *a political problem*. This long term transformation may be described as the 'democratisation' of fear, not in the ridiculous sense that everyone comes to exercise their right to be afraid, or is duty-bound to be so, but rather that fear, especially its debilitating and anti-democratic forms, ceases to be seen as 'natural' and comes instead to be understood as a *contingent* human experience, *as a publicly treatable phenomenon, as a political problem for which tried and tested political remedies may be found*. (Keane 2001: 32–35; emphasis added)

Keane is absolutely right in reminding the subjects, analysts and even the practitioners of various kinds of fear – including those social scientists who are 'fearful' of being misunderstood as siding with the climate skeptics and therefore shy away from dissecting the dichotomy of 'climate fear' – that, 'the political effort to identify fears, to name them, to witness and care for their victims, and to hunt down their perpetrators so that they might be brought before courts of law, is something positive, yet incomplete' (ibid.: 35). Moreover, concludes Keane, 'It is hard to know where today's democracies are positioned on the scale of either understanding the fears that they (or other regimes) generate, or their counter-capacity to cultivate fearlessness, for instance through publicly witnessing the dastardly effects of fear' (ibid.).

If we were to glean from the various insights offered by Keane the proverbial billion dollar question for the purposes of critically examining the 'new' but nuanced category of 'climate fear/terror', it might be this: against the backdrop of the growing 'democratization' of 'globalized' fears, how do we distinguish and define the concept of 'climate fear', demarcate its spaces, and identify its 'objective circumstances', 'subjective systems' and 'abreactions'?

Needless to say this question begs insights from various other disciplines, especially geography. The new geopolitics of fear, argues Rachel

Pain (2009) is 'globalized' in the sense that 'emotions are positioned as primarily being produced and circulating on a global scale, rather than rooted in the existing biographies of places and their social relations; and second, in that they tend to be discussed as though they apply to everyone all of the time. Globalized fear is a 'metanarrative that tends to constitute fear as omnipresent and connected, yet at the same time analyses it remotely, lacking grounding, embodiment or emotion' (Pain 2009: 467). She would argue that there is a significant geography to the so-called 'globalized' fear in the sense that 'Global risks and threats do not map neatly onto local fears' (Pain 2010) and moreover 'those most affected by fear in the current geopolitical climate are marginalized minority groups' (ibid.). Our analysis of the new geopolitics of climate fear in the chapters to follow endorses these insights.

A critical geographical-politics of 'climate terror' cannot afford to downplay the trends that suggest a strong political push by the nascent transnational coalition devoted to – and hoping to benefit from – scientific-technological innovations in the green industry, toward frightening the masses – especially the marginalized and the vulnerable – *away from* politics and the political. To note an example in passing, the first national adaptation plan of the United Kingdom, announced on 1 July 2013, describes climate change as a 'big business opportunity', claims global leadership for the UK in this industry and market for the future, and reassures the UK companies of full government support (United Kingdom 2013).

What appears to be common to both the war on terror and the securitization/militarization of climate change is the speculative preemption of future threats and dangers to justify the manipulation of socio-spatial consciousness and policy interventions by the powers that be in the name of a moral economy that is heavily skewed in favor of the securing of the future citizen. Challenging the accumulative terrorizing effects of climate fears is not an easy task since what is at stake are the interests and agendas of various state and non-state actors.

Describing various facets of the 'non-political politics of climate change', Erik Swyngedouw (2013) points out a 'paradoxical situation whereby the environment is politically mobilized, yet this political concern with the environment, as presently articulated, is argued to suspend the proper political dimension.' He underlines the need to further explore 'how the elevation of the environment to a public concern is both a marker of and constituent force in the production of de-politicization' (ibid.: 2–3). Ulrich Beck, in agreement with this contention would point out that by citing certain 'indisputable facts' and predicting a catastrophic future for humanity, 'green politics has succeeded in de-politicizing political passions to the point of leaving citizens nothing but gloomy asceticism, a terror of violating nature and an indifference towards the modernization of modernity' (Beck 2010: 263).

Swyngedouw (2013: 3) would go on to argue that notwithstanding the fact that Nature continues to be mean different things to different individuals and social groups embedded in diverse cultural settings, we find an overwhelming consensus 'over the seriousness of the environmental condition and the precariousness of our socio-ecological predicament'. Few, in his view, would dispute the imperatives of environmental sustainability if disasters were to be averted. According to Swyngedouw,

> In this consensual setting, environmental problems are generally staged as universally threatening to the survival of humankind and sustained by what Mike Davis (1999) called 'ecologies of fear' on the one hand and a series of decidedly populist gestures on the other. The discursive matrix through which the contemporary meaning of the environmental condition is woven is one quilted by the invocation of fear and danger, and the specter of ecological annihilation or at least seriously distressed socio-ecological conditions for many people in the near future. (Swyngedouw 2013: 3)

The politics of non-political and the post-political is no doubt gaining momentum but not without inviting significant resistance, emanating from a growing awareness of unequal geographies of neo-liberal globalization in various parts of the globe, and articulated through trans-local networks. We agree with David Featherstone that a politics that fails to question hierarchical power relations related to climate change runs the risk of becoming redundant. Engaged in this counter-politics directed at unequal power relations and nation-state centric narratives of the political which continue to dictate the debate on the 'post-political' are equally innovative 'alliances and exchanges between different groups and activists based in different parts of the world' (Featherstone 2013: 50). Featherstone further argues that scholars like Erik Swyngedouw tend 'to ignore the ways in which contestation to climate change, such as the organizing in advance of COP15 and the alliances configured through the protests, exceed, unsettle and undermine attempts to contain contestation within the nation. These mobilizations suggest that forms of contentious politics shape more generative geographies of antagonism and more diverse modalities of contestation than is acknowledged by theorists of the post political' (Featherstone 2013: 47).

In some ways 'climate fear' is similar to other forms of fear (including 'globalized' and 'everyday') and yet distinct in several ways. As Pain and Smith (2008) have pointed out, 'it is more useful to see fear as made up of a range of multiscalar influences constituting an assemblage, rather than assuming the spatial hierarchy where global processes have local impacts on feelings.' The category of climate fear also appears to defy both spatial and temporal compartmentalization. As an assemblage of geopolitical, geoeconomic and strategic-military agencies, agendas and policies – anchored in earth climate science – 'climate fear' draws, on the one hand, on the anxieties emanating from natural-social-political disasters of various magnitudes and scales. On the other hand, its roots can be traced back to the geographical politics of 'nuclear winter' during the cold period. Before we attempt to briefly map various *routes* that fear-terror inducing climate change discourses have taken, a brief reflection on some of its *roots* could be highly illuminating.

As the industrialized nations and the vested interest groups, having lead the globalization project, struggle to put a gloss over the *causes* of climate change by invoking the catastrophic *consequences* of the so-called 'global' challenge, the 'strategic' deployment of old metaphors to frame the new geopolitics of climate terror has assumed intriguing features.

'Environmental possibilism', about which the Sprouts wrote in the late 1950s (Sprout and Sprout 1956), stands profoundly challenged by 'climate determinism'. What the Sprouts were suggesting was the need to map out how the milieu can enhance/facilitate or constrain/inhibit the ability of a state to act. In contrast, so profound and far-reaching is the climate metanarrative with its offshoot master discourses that the external, invisible but omnipresent hand of 'unitary agency of climate change' is being seen as exercising a determining influence on almost each and every domain that concerns humans activities and intercourse. What is the core essence of climate change? Is it an environmental-ecological issue? Is it about nature and uses of energy resources? Is it a threat to national security? Is it a human security issue? Some of these or all of these?

Communicating the imaginative geographies of climate fear: promise and pitfalls

The critical take of Saffron O'Neill and Sophie Nicholson-Cole (2009) on fear-inducing or shock-provoking representations of climate change is that despite some promise for drawing popular attention to climate change, fear appears to be neither an appropriate nor an effective catalyst for invoking a 'genuine' personal response to and engagement with climate change. In their view, images, metaphors and icons that are 'non-threatening in nature' and relate to everyday emotions and concerns of 'individuals' in the context of this macro-environmental issue tend to be the most engaging' (ibid.: 355). Since there is a highly complex geography to both the materiality and perceptions of 'risks' and 'catastrophes' – the prevailing and the probable – it is rather difficult to find framings, howsoever dramatic, that could act as common climate rallying points for the US citizens in Manhattan, the slum dwellers of Mumbai and the coastal communities of Somalia. For millions situated on the margins, especially in the global South, whereas the roots of 'climate catastrophes' are intricately entangled with deeply entrenched and disempowering social-economic hierarchies, the routes are crisscrossed by trajectories of ecological injustices and acute livelihood insecurities. In a sharp contrast to the anxieties and fears of the affluent urban dwellers, the relevance of climate change is far more multifaceted, multilayered, and much more nuanced for the subaltern.

As Delf Rothe (2011: 330) puts it so aptly, 'The growing consensus on dangerous climate change ultimately reinforces this advanced liberal risk management by presenting climate change as a 'naturalized' de-bounded risk and blurring its socio-economic causes.' The media 'stories' of global warming and climate, consciously or unconsciously, are implicated in and also contribute to 'manufacturing consensus' around uncritically accepted virtues of neoliberal, technocratic-managerial approach and solutions to climate change.

Climate terror and the 'new wars': militarization of climate change

There appears to be a growing militarization of climate change and the concomitant up scaling of 'climate fear' to 'climate terror' on yet another site of vital importance; military-defense establishments. We are inclined to agree with a reasonable fear expressed by some keen analysts that terrorism and climate change might become inextricably linked in the mental maps of the foreign policy establishment and the general public, as certain influential think tanks and their analysts craft highly alarmist scenarios in which climate change-induced political instability is used as a pretext to continue the war on terror (Hopkins 2008). In certain parts of the global North, imaginative geographies of political instability, displacements, conflicts, migrations 'induced' and/or 'multiplied' by climate change, are giving rise to defense-military-security scenarios that 'naturally' feed into the 'war on terror' and its legal-political exceptionalism.

For example, a recently established think tank named 'Center for a New American Security' (CNAS) while talking about the impacts of climate change makes as many as 37 references to terror, terrorist, or terrorism (Hopkins 2008). According to the Report, 'While we continue our debates and disagreements, wouldn't it be wise to take steps, particularly when many of them are financially attractive, which reduce both the risk of mass terrorism and the chance of catastrophic climate change?' (cited in ibid.).

Speaking of 'mass terrorism' and 'catastrophic climate change' in the same breath shows how a causal relationship is being uncritically established between these two terrorizing abstractions. A number of studies have appeared in the course of the past decade and more focusing on national security implications of climate change, prompted in part perhaps by a growing acknowledgment of the vast complexity of climate change by the powers that be, on the one hand, and by the nagging absence of effective mitigation and adaptation measures backed up by the necessary political will. It is important to note the convergence between the geo-economic and geopolitical-strategic reasonings of national security.

One of the early reports that received large-scale circulation and attention was authored in 2003 by Peter Schwartz, a CIA consultant and former head of planning at Royal Dutch/Shell Group, and Doug Randall of the California-based Global Business Network, and entitled 'An Abrupt Climate Change Scenario and the Implications for United States Security' (Schwartz and Randall 2003). While daring themselves with the task of 'imagining the unthinkable', the authors argued that the sudden onset of climate change would be little short of catastrophic, and pronounced in a Hobbesian sense: 'Disruption and conflict will be endemic features of life ... Once again, warfare would define human life' (ibid.: 17). Arguing that the enormity of the risk inherent in such a scenario was so huge, Schwartz and Randall wrote that climate change needed to 'be elevated beyond a scientific debate to a U.S. national security concern' (ibid.: 3).

One of the most alarmist features of this report was its contention that, 'abrupt climate change is likely to stretch carrying capacity well beyond its already precarious limits' and 'as abrupt climate change lowers the world's carrying capacity aggressive wars are likely to be fought over food, water and energy' (ibid.: 15). The authors of the report however tried to mitigate the dreadful sounding consequences of the eroding 'carrying capacity' by climate change by saying that, 'Deaths from war as well as starvation and disease will decrease population size, which overtime, will re-balance with carrying capacity' (ibid.). The imprint of Malthusian thinking was unmistakable on the report, which offered a new discursive map of the globe in terms of the so-called 'carrying capacity'. We are further told that, both at the regional and state levels, countries situated in global North, endowed with high carrying capacity and resource-population equation happily tilted in favor of the latter, are likely to adapt far more effectively to abrupt climate change. Consequently, 'this may give rise to a more severe have, have-not mentality, causing resentment towards those nations with a higher carrying capacity' (ibid.: 16).

This report, flagging its task of 'imagining the unthinkable', laid down the foundations of a trend that has become more popular and pronounced with the passage of time: growing fascination with fear-inducing, diverse strands of 'anticipatory regimes' with regard to climate change. One such trend weaves in a neo-Malthusian climate conflict narrative, which is fed by, and feeds, the fears of overpopulation in the global South, casting a perennial dark shadow on earth's 'common' carrying capacity. Here is a complex geopolitics of anticipation with far-reaching ethical considerations, which we take up in the concluding chapter of this book. Suffice to note for the time being that, as Adams, Murphy and Clarke describe (2009: 249):

> Anticipatory regimes offer a future that may or may not arrive, is always uncertain and yet is necessarily coming and so therefore always demanding a response ... Anticipation is not just betting on the future; it is a moral economy in which the future sets the conditions of possibility for action in the present, in which the future is inhabited in the present. Through anticipation, the future arrives as already formed in the present, as if the emergency has already happened.

The greater the fear on the part of military-defense establishments in the global North of an impending catastrophic climate change, the greater is the inclination toward seeking solutions for future climate problems through military intervention.

Will imaginative geographies of climate-induced disasters, displacements and devastation create grounds for 'new' humanitarian interventions by the national security-defense establishments, especially in the global North? Jim Thomas, a member of the Centre for New American Security (CNAS) advisory board, argues that, 'At times, the United States may have to use its military to prevent wars and advance its interests, not simply to "fight and win wars" when they occur. Accordingly, the application of the military should in some cases be not of last resort, but should sometimes occur at an early date when it is still possible to prevent security problems from metastasizing and affecting the broader international security system, and U.S. interests more directly' (Thomas 2008: 7). In his view: 'Such a preventive approach would represent a departure from the so-called Weinberger doctrine on the use of the military – as a last resort and only when vital interests are threatened – and a reorientation toward earlier involvement in problem areas by working with and through others to address security challenges so that large-scale intervention by the United States is less likely at a later time' (ibid.). How the military establishments, situated in various political-strategic cultures, are going to put into place anticipatory regimes for addressing climate change induced threats is far from being clear at present.

As pointed out by John Wihbey (2008) these days one finds a growing deployment of military metaphors in environmental contexts, such as a 'Manhattan Project' for clean energy, a 'Marshall Plan' for green action and so on. The Pentagon too has been looking at long-term, low probability events marked by an alarmingly high level of probability such as weapons proliferation and incremental climate change. Were this to happen with regard to various parts of Afro-Asia, for example, in 'anticipation' of climate anarchy, the US military-intelligence operations will be seen as another extension of 'continued belligerent interventionism' on the pretext of apparent 'humanitarianism' but with deeply geopolitical motives (ibid.).

Conclusion: crisis of statecraft, neoliberal regulations and climate governance – leaking geographies of sovereignty

One of the key climate change puzzles, partly geographic and partly political in nature, relates to what Delf Rothe (2011: 330) has rightly described as a 'paradoxical moment in world risk society'. How is it that despite the IPCC 'consensus' – arrived at by both earth climate scientists and the relevant bureaucracies of the countries they represent – the response of various state parties, (despite rhetorical commitments shown and shared so enthusiastically at the Conference of Parties to climate change mitigation and adaptation) is so varied in terms of domestic responses and implementation?

In our view it is equally useful to note that while climate is changing at a pace and scale that is yet to be figured out comprehensively in terms of implications for the entire globe, the 'nature' of Westphalian sovereign states appears to be changing in the wake of globalization and its regional geoeconomic-geopolitical manifestations, including various free trade agreements. Whereas it is neither feasible nor desirable to generalize about the nature, scope and implications of ongoing transformations in the 'stateness' of nearly two hundred countries, pitched currently at different levels of uneven development, it is possible to argue that these diversely situated nation-states are experiencing serious challenges – in many cases bordering crisis – to their authority, legitimacy and governance in the wake of globalization.

The link between climate crisis and crisis of statecraft (which is reflected in but not subsumed by the 'crisis of governance') is important, we believe. It demands an analysis of how climate discourse is being up scaled by the sovereign state actors in the light of the 'climate of fear' caused by the de-territorializing mobility of various kinds of boundary-defying networks, and the sharp decline in the capacity to govern with sufficient authority and legitimacy. One of the key challenges before the modern states with withering, leaking state-centric geographies of security and sovereignty is to stage a comeback, despite growing odds posed by capital-material, human and communication flows. In the 'global risk society' constituted by climate change, scarcities and neoliberal globalization, innovative alliances, and alignments of the 'like-minded' are in the making, and carry sufficient promise to reconfigure 'global' geopolitics in the 'Asian' century.

The 'nature' turned/transformed into 'environment' (and now overwhelmed by 'climate') is no longer 'external' to state and statecraft – something to which a state would respond in times of 'natural disasters' with its military or paramilitary forces. It is integral to, and indistinguishable from, statecraft and its crisis. Moreover the politically inconvenient juxtaposition between the 'end of nature' and the 'crisis of statecraft' has in some (not all) cases led to situations in both global North and global South where the 'fury of nature' or 'natural disaster' are seen as some of the greatest challenges to 'national security,' and not as concerns of human insecurities caused by crisis of governance.

2
Climate 'Science': Categories, Cultures and Contestations

Introduction

It is not our intention here, as stated clearly and necessarily in the opening pages of this book, to pit climate change deniers against climate change advocates; it is not our objective to argue who is *right*. Rather, we are willing to accept that the weight of evidence seems to be heavily in favor of those scientists who *do* believe that human-induced climate change will become an increasing feature of our global physical and social environments. As social scientists, however, it is not sufficient to simply make a judgment in defense of those advocating a particular scientific position. Rather, we are interested in what kind of 'science' (knowledge in all its forms) is being articulated; why it is being pursued; *which* scientists are advocating these positions; and, what are the geopolitical outcomes of these scientific findings and solicitations. Science is largely a human construct. In this book, however, we acknowledge that there also exist essential 'natural', non-human forces, which are larger than the agency of homo-sapiens can allow for, imagine, or control. But where the human agency/essentialist nature balance becomes problematic is when understandings of an 'essentialist nature' are used to empower certain minority world communities to the detriment of others living and surviving in majority worlds. These divisions in power based on elite knowledge are reproduced in geopolitical maps, privileging certain parts of the globe, prioritizing the development of certain communities over others. We need to know where these divisions and boundaries are, and how climate change is deterritorializing the globe, based, in large part, on climate science.

Science and environmentalisms

Before we ask these questions, however, we must understand the unique (and often turbulent) relationship that the modern environmental movement has had with western science since its first emergence in the mid- to late-1960s, and how, in turn, this relationship has been rearranged in more recent times. The climate change movement must be understood in this work to be only a recent incarnation of environmental or green movements. It is, in actual fact, a sub-set, or a series of networks of environmental activists, which have come together, particularly since the 1990s, to elevate climate change to the most widely advocated of all environmental issues. So powerful have these climate networks within the broader movement become, that many believe that all environmental issues can now exclusively be understood within its rubric or frame. Climate has become the *omniscient* environmental issue of the 21st century, to such an extent that other issues have been sucked into its conceptual vortex, usurped by it, trumped by it.

The relationship between traditional western science and the early modern environmental movement was always ambiguous, but was largely an antagonistic one. In many ways, early modern environmentalism defined itself in rugged opposition to western science. Western science was deemed a central plank in the project of modernity, and early environmentalists vociferously challenged this project, with its focus on progress at-all-costs and its pursuit of technological, reductionist science-driven solutions to the earth's problems. In fact, Western science, with its knowledge egocentricity, was seen as a major cause of many environmental problems on the planet.

From this view, science was regarded as an integral factor in determining a 'Western', derogatory view of nature. Warwick Fox, in his book *Towards a Transpersonal Ecology* (1990), lists six characteristics of what he terms as 'unrestrained exploitation and expansionism'. They can be summarized as follows:

1. It emphasizes the value to humans that can be acquired by physically transforming the non-human world.
2. It not only measures the physical transformation value of the non-human in terms of economic value, but also tends to equate the physical transformation 'resources' with economic growth. Economic growth is then equated with progress, which is seen as an unquestioned pathway for future human endeavor.

3. It advocates the 'Myth of Superabundance' – legitimizing continuous expansion – adopting a frontier or cowboy ethics; an ethics of 'How the West was won'.
4. It is totally anthropocentric; the non-human world is considered to be valuable only insofar as it is of economic value to humans.
5. It is characterized by short-term thinking, that is, its anthropocentrism does not include the interests of future generations.
6. When forced to consider longer term (deleterious effects of continued unrestrained exploitation and expansionism) – or continued 'business as usual' – this approach falls back on technological optimism (adapted from Doyle 2001: 11).

This list of attributes is an excellent shorthand version of what one of us has termed as *unrestrained use* (Doyle 2001). This view of the world, with its clear dichotomy between humanity and the 'rest of nature' has its origins in ancient Greece and Judeo-Christian belief systems. It was then coupled with the advances in knowledge and technological advances derived from the scientific and industrial revolutions. This dichotomy, or profound dualism – which only saw value being attributed to nature when it could be transmogrified through 'human use', was understood by these early environmentalists to have led to extreme planetary degradation, often blinded, as it was, by technological optimism.

A key component of this post-Enlightenment world-view was that scientists (both 'natural' and social) assumed key roles in promoting the all-embracing concept of progress. In the 'ancient' view of the western world, the predominant view was that the 'golden age' existed in the past somewhere, some place; but the new men of science, ably supported by philosophers such as Kant and Saint-Simon (Bury 1960), now depicted the golden age as lying in the future. Krishnan (1978: 14) argues that humanity could now be understood to be advancing, 'slowly perhaps but inevitably and indefinitely, in a desirable direction'. In a sense, it was illogical even to attempt to comprehend the end-point of this progression; but the attraction 'proved irresistible'. Krishnan goes on to argue:

> The future beckoned urgently, and the promise it held out could only adequately be gauged by the chaos that might result if the forces of progress were not all combined in the task of bringing the new society into being. Of those forces the most important were science, the men of science, and all those who could see in the achievements of

> the scientific method the highest fulfillment of the Enlightenment, and the key to the future direction and organization of society. (Krishnan 1978: 26)

Progress, as a concept, became a dangerous myth in the context of environmental crises when, as Bury (1960) writes, it was linked with the Baconian idea that scientific knowledge should be applied to human manipulation of the natural environment. Progress came to be seen as intrinsically tied to changing, taming, and controlling the non-human world. Nature could now be scientifically and technologically 'improved' upon, and many environmentally degrading acts ensued under the banner of 'western progress' (Doyle 2001: 112).

We say 'western progress' quite deliberately here, as this central underlying myth of modernity was rudely ethnocentric. Modern (Western) humanity had produced cold reason and science. Supporters of progress mused: 'What were those "primitive" societies in comparison with this great intellectual, political and social wealth?' This was very much the belief system that took theories of progress and modernity into the mass domain after World War Two. Modernization Theory, based on this very same Western scientific reasoning, became the key driving ideology of Western developmentalism during this post-war period and, although it has suffered serious bouts of attack from critical thinkers of many persuasions, we would argue that it remains central to the dominant western (and now global) paradigm regarding human and environmental development.

Modernization advocates argued that, now that both social and 'normal' scientists possessed *the* historical model – the modernization of the West –, they were able to observe and measure this process (Greig et al. 2007). In turn, this model could be viewed as a successful prototype that the object of enquiry – the poorer people of global South – could imitate. Also, while the 'original transition' to modernity was primarily viewed as lying firmly at the feet of internal factors in 'problem states' (such as addressing corruption, internal political systems etc.), this scientific modernist prototype helped to promote development and modernization through assistance from the West. Other knowledge systems of the non-western world became categorized as myths, as falsehoods, usually articulated through 'quaint' symbols and rituals, whilst the West had finally attained, if not the absolute truth, then the right theoretical and experiential tools with which to pursue it.

What was now particularly needed was for the less affluent world to adopt all manner of western behaviors and customs, which had been determined by science as superior. The preferred developmental attributes included: Western economic and political systems, language, education, gender relations, religion, sports, and the list goes on.

As aforementioned, critics of post-war modernization theory have been many, including scholars and activists coming from dependency theory and world systems backgrounds. Most of these critiques are based on understanding the source of economic disparities between those nations at the *core* of modernization, versus those on the *periphery*. Indeed, these arguments will be picked up again in Chapter 4 on climate markets and economics. All that needs to be mentioned here is the general tenor of this opposition to Western science-led modernization. First of all, and most obviously, profound questions exist as to whether empirical observations made throughout Western history can be utilized firstly to comprehend, and then to scientifically predict, developmental futures into 'other cultures' (Greig et al. 2007). In this vein, this model for development – rather than a global model – has been roundly criticized as soliciting and promoting both westernization, ethnocentrism and, on occasions, quite simply racism (Rodney 1972).

Along with scientific method being elevated to the plane of essential truths, Western economic and political systems such as capitalism and pluralist democracy (supposedly built on superior western reason), were also seen as a cut above all else, and in desperate need of being exported to *outside* civilizations. An even more basic objection to the premise that global inequality could be successfully overcome if Western ideas and values were transported to the peripheries was that the modernization project ignored the fact that post-war societies in the global North were far from perfect, with enormous cleavages still existing within them based on inequities defined by gender, class, race etc.

So, for many in early green movements, western science-led modernization – fuelling imperialism, capitalism and other industry-intensive regimes – was seen as *the* major contributor to both global environmental degradation and for creating enormous disparities in wealth and access to resources between the rich and the poor. The divisions been global haves and have-nots had been carved out across the peoples of the planet on these principles and beliefs. Indeed, the short-hand division which we often use in this work between the global North and the global South was forged upon these massive discrepancies in access to, and ownership of, these elite knowledge, production and consumption systems. These early environmentalists fought, in both terms of ideology and the kinds of political repertoires they utilized, in rugged and direct opposition to this dominant form of Western-style development.

They argued for absolute limits-to-growth, articulating the concept of a finite planet (see Chapter 4). They also firmly believed that the Earth was more precious, more complicated, and more profound than limited Western science could value and account for.

Like all social movements, early days are often spent fighting on the 'outside' of dominant political institutions and their paradigms of knowledge. As the 60s moved into the 70s, some criticisms against more reductionist and elite Western science were *accommodated* for by more radical eco-philosophers and scientists. This was not an attempt to cease demonizing modernist science, but rather build into it functions which could cope with the more strident forms of oppositional criticism against development interests emanating from adversarial environmentalists. The work of Barry Commoner is of interest here. His 'Four Laws of Ecology': 'everything is connected to everything else; everything must go somewhere; nature knows best; and, there's no such thing as a free lunch' (Commoner 1971), are neat maxims based on the 'new' science of ecology, and achieved some prominence within Western green movements. The first two and the fourth laws were really part of one idea: interconnectedness, while the third law constituted a myth in every sense, as it gave ultimate knowledge to a powerful force that is beyond human science and reason: both the Gods of science and the western theists had been partially replaced with Nature with a capital 'N'.

As the 70s moved into the 80s, other more complete ways of accommodating the ideology, the politics of environmental concern – and its opposition to Western, science-led developmentalism – began to emerge in the global North. Supporters of the modernist project oversaw its clever rebuild, a renovation which, in the twinkle of an eye, saw the Project emerge not as Nature's enemy but as its savior. And so the language of *ecological modernization* was born, and with it, its close cousin sustainable development (see Chapter 4).

Science and ecological modernization

Early environmentalists, in understanding the Earth as finite, saw environment versus development tussles in the context of zero-sum games, with win-loss outcomes. The brilliance of the ecological modernists was to develop an approach to resolving developmental and environmental objectives by depicting them as achievable through a positive sum game. Heavily reliant on scientific expertize and technology-driven responses, ecological modernization encapsulated a discourse which was market friendly, working in close collaboration with development and business interests. As will be discussed in Chapter 4, it was now argued that economic growth and environmental protection were now both coherent with *sustainable development*.

Adherents of the theory basically challenged the view that economic growth must be accompanied by ecological damage. More mainstream theorists like Mol and Spaargaren (2000), referred to disparagingly by Christoff (1996) as proponents of the 'weak' form of EM, focused on more economistic, national, top down approaches: heavy on technological solutions; and on the formation and collaboration of policy elites. The emphasis was now on innovation, flexibility in science and management, cost efficiency, market-driven solutions, and collaboration between business and government. Whilst derived from the more corporatist European states (for example, Denmark and Germany), this dominant supposedly 'weaker' strand quickly became the predominant interpretation, and the US quickly morphed into one of its main discursive agents (see Chapter 4 for EU emissions trading and Cap and Trade as examples of this).

In a positive spin, writers such as Mol (2003) claimed that, in fact, ecological ideas were reshaping the central institutions of our society, including government and business. On the principle that being green makes economic sense, societies and economies were being transformed by ecological thinking. Mol's book (2000), *The Ecological Modernization of the Global Economy*, aimed to demonstrate how, in fact, globalization was facilitating ecological reform and thus facilitating the ecological modernization of the global economy. Hence, Mol argued, what was emerging was not just a western or industrialized economy-type phenomenon, but a greater 'harmonization' of international standards benefitting many different countries. Where it did not occur, Mol argued, in classical modernist fashion, that this was the result of local policy or endogenous political failures of the periphery.

To be fair, not all ecological modernists were so all-embracing of the traditional modernist line, but did see some advantages in this newer, less adversarial and less conflict-driven version of environmentalism. For example, writers such as Hajer (1995) and Christoff (1996) argued for more bottom-up versions of EM, and that social movements must be more actively engaged in decision-making (they often labeled themselves flatteringly as proponents of the 'strong' understanding of EM). Their focus was more on the discursive elements of EM, justice and democratic participation in policy-making; going beyond just better and more efficient housekeeping; and implying the substantial restructuring of the economy and industrial society.

Whilst acknowledging that ecological modernization and sustainable development can be interpreted in various and often ambiguous ways, Gibbs (2000: 11) concludes that ecological modernization – at least in its 'weak' form – has significant problems. He writes:

> This enabling state will deliver ecological modernisation through corporatist relationships between government and industry, although co-opting environmental movements where necessary, thus ignoring issues of participation and reducing the rest of society to passive consumers to be provided with enough information to make informed (but market-based) choices. (ibid.: 11)

Regardless of whether ecological modernists existed at the 'weaker' or 'stronger' end of the ideological spectrum, there can be no doubt that the more pliable, 'weaker' version of ecological modernization achieved the upper hand during the late 1980s and grew more dominant throughout the 90s. Like the classical modernist position, writers such as Mol used Western world examples of 'successful' sustainable development as a scientific and technocratic blueprint upon which the rest of the world must aspire and build upon. In a more critical work on ecological modernization, Toke argued that Mol's classic study of the Dutch chemical industry is not predictive of what might happen in other sectors and industries and, as a consequence, its success cannot be safely predicted and cannot be used as a normative approach (Toke 2011a: 12–20; also Toke 2011b).

Another excellent critique of ecological modernization and its sister concept, sustainable development, emerged in the writings of John Bellamy Foster (2009, 2012). He argued that the modernization theory of the post-WWII period had not been debunked, but paid lip service to the needs to be more 'reflexive' in attitudes to development. EM focuses on the processes of production, but not the production relationships themselves (Foster 2012: 213). The 'close family resemblance' of EM to earlier modernization theory is, he argues, hard to miss (ibid.: 216). EM, regardless of its more recent ideological jockeying, because of 'its teleological commitment to progress, its acquiescence to the status quo, and its lack of attention to the larger sociological, economic, and ecological context', (ibid.: 225) was still really just an extension of the classical modernist project. Bellamy Foster considers EM to be deeply flawed, both empirically and theoretically. EM advocates like Mol do not provide evidence of the actual effectiveness of the initiatives they defend, because they focus on the micro – the more specific, localized success stories – and do not address issues on the macro level effects. He accuses EM supporters – both 'weak' and 'strong' – of questionable empirical methodologies (ibid.: 221–222). Furthermore, he lambasts EM supporters as taking a skeptical, postmodern view of science, which allows them to ignore the scientific evidence to suggest the growing environmental problems and the lack of evidence to indicate that EM approaches are able to mitigate these. They are guilty of modernization 'exemptionalism' (ibid.: 223).

EM ignores issues of power entirely, ignoring fundamental classical sociological variables, and the control over science and technology by elites:

> The emphasis on authority (increasingly private authority) as opposed to power or hegemony is thus a product of ecological modernization's focus on elite policy-driven processes rather than social struggle. The result, however, is the systematic marginalization of key sociological variables related to environmental justice: class, gender, race, international exploitation, and even democratic movements and struggles. (Foster 2012: 224)

So, as the latter part of the 20th century moved on, the more radical and more oppositional messages of environmental movements in the global North were being sidelined. Interestingly, this was the very time in which the environmental movements began to get traction in the global South. In its nascent days, the predominantly Northern environmental movement (despite its early opposition to science) was still largely seen as an off-shoot of the white, first world, with third-world thinkers and activists remaining very suspicious of its knowledge systems being transported into their midst. Obviously, the linking of environmental and developmental problems through the rhetoric of EM and SD brought environmental problems within the more usual realm of politics and economics in third world countries. But also, problematically, environmental problems in the North were still cast within both post-material and post-industrial frameworks; whereas in the global South, societies were seen as mostly *not* post-material, *not* post-industrial, so the ecological modernization goals and programs made little sense. Worse than this, ecological modernization was often seen as a form of green imperialism, even by Southern environmentalists, who were pursuing green goals through a post-colonialist frame (these ideological distinctions between green emancipatory positions is taken up in some detail in Chapter 7).

Climate change: The *ideal type* of ecological modernization

It is within this historical background that we now introduce the environmental issue of climate change. Climate change, as a symbolic flag, towers over all other environmental issues in the opening stanzas of the 21st century. Climate change is the Weberian *ideal type* of environmental issue emerging from discourses of ecological modernization and sustainable development. Without the transition from early oppositional environmental politics to more corporatist forms of politics, which produced EM and SD, climate change, as an issue, would not have achieved such prominence, regardless of any *essential* importance, crisis imperative, or Earthly urgency attached to it. In Chapter 4, we look at the way in which neo-liberalism (a key component of environmental modernization) embraced the issue of climate change. Similarly, in Chapter 6, we show how climate change emerged as the cause célèbre of all environmental security issues – again, serving the ecological modernists debates about 'securing' the planet Earth. But what largely remains to be done in this chapter is to explain how climate change emerged as the ultimate ecological modernist issue, through its science-centeredness.

The work of David Demeritt (2001) is particularly interesting in discussing the key components of climate change that reinforce its inherent links with reductionist, modernist science. Much of his work here relates to the role of General Circulation Models (GCMs) (the dominant means of climate analysis) and is one of the central planks of his argument. He analyses the history of climate modeling: GCMs are dominant over other modalities of climate science, establishing a research hierarchy privileging the physical sciences over the life and social sciences. Some of his criticisms of scientific modeling include:

- Climate models are physically reductionist. They consider only the physical properties of GH gases and all human social contexts are divorced from the analysis and left to policymakers and politicians. The essential question is only the gross emissions, not the reasons why the emissions are made. Unequal power relations underlying this are ignored, the existing power relations thus essentially maintained.

- Models are premised on assumption that complex systems can be broken into their constituent parts from which modeling can take place from the bottom up. But that's not how it actually happens, '... many underlying physical processes are parameterized' (ibid.: 317). Hence, GCM science is given an exaggerated predictive capacity, which the models are really not able to provide. Also, the models are based on a 'business as usual' model, which hides possible responses to the problem, i.e. the west could consider not following 'business as usual' (ibid.: 318).
- The construction of modelling science responds to scientists' tacit beliefs that they must provide predictive science to feed the 'practical' needs of policymakers downstream. One such case is the perceived need to provide modelling to support and develop regional responses to climate change, leading scientists to respond to this perceived need by using modelling, even though ...

  Arguably, the application of complex GCMs to the generation of regional climate change scenarios for impact assessment is tantamount to using a laser guided missile to swat a fly: a fly swatter might do just as well without suggesting a degree of precision unwarranted by unknown levels of modelling uncertainty and future indeterminacy. (ibid.: 319–320)

As touched upon in the opening pages of this chapter, despite the fact that, on the whole, many early environmentalists were deeply suspicious of science and technology, the relationship has always been an ambiguous one. Although Western science, through ecological modernization, has now captured most of the modern environmental movement's endeavors, like any social movement there remain outlying networks who maintain their attack on science.

In the current case of global food sovereignty campaigns (Sommerville et al., 2014), many environmentalists still frame themselves as largely in opposition to science. They question the Western scientific paradigm, in a Kuhnian sense (Kuhn 1961), and argue that limits must be placed on scientific and technological programs – the results of these pursuits threatening the fabric of the earth, its universe. Ultimately, science – unmediated by the state and civil society – produces inequitable outcomes for the majority of those who inhabit the planet, both human and non-human. Science, in this vein, is seen as unquestionably pursuing the gains of Northern elites, another wave of imperialism; a primary value-system or world view which colonizes the planet in much the same way as Christian religions have done before it.

The position of many environmentalists in relation to dominant paradigmatic science is just as intense in the current case of climate change, but usually occupies a position on the opposite side-of-the fence. On this occasion, green activists not only make scientists bedfellows but, in fact, champion scientists. Many green activists support the now nearly universal call from western science that the planet is warming, due largely to human pressures. What makes this case fascinating is that the issue exists purely within scientific data, working in millennial scales; it is not *experienced* as such in human lifetimes. The issue would not exist if it had not been summoned by Western science. This conflicting relationship to science itself is an integral component of campaign framing, and has a profound impact upon repertoires experienced in environmental campaigns (Doherty and Doyle 2014).

Of course, the answer simply may be that green groups make a strategic coalition with scientists when it suits their case (rather than being necessarily convinced of the science per se), as they do on a daily basis with other attempts at coalition and network building. But there is no doubting that the power of climate change as a campaign focus right across the globe is now without comparison and, in large part, this is a result of the current dominance of the ecological modernist frame. Of all current environmental issues, climate change has shown a remarkable ability to reach out across groups, touching most, and being *the* symbolic banner under which so many diverse forms of environmental actions take place. In a biophysical sense, climate – like the wind – crosses boundaries, moving in, through, over and under the politics of nation-states. Climate change, it could be argued, is the ultimate post-cold war, post-modern issue. It comprises a truly transnational (but predominantly western European and North American) theme, and was further remodeled in the first euphoric waves of multilateralism and an emergent cosmopolitanism taking root in the decades either side of beginning of the new millennium. It would seem obvious that such an issue would be embraced enthusiastically by transnational green organizations, struggling at times to move beyond their historic roots in the North to become truly global organizations. Of course, there are also advantages in taking part in these identity-building pursuits by more peripheral actors operating in the global South. Atkinson and Scurrah (2009) state: 'In this process network relationships are key. For the members in the South the network provides access, leverage and information. For those in the North the linkages with the Southern members provides credibility and a mandate.' As Mary Kaldor (2013: 95) says of these transnational civic networks:

> They represent a kind of two-way street between Southern groups and individuals, or rather the groups and individuals who directly represent victims ... with the so-called Northern solidaristic 'outsiders'. The former provide testimony, stories and information about their situation and they confer legitimacy on those who campaign on their behalf. The latter provide access to global institutions, funders or global media as well as 'interpretations' more suited to the global context.

Northern environmentalists seek not only credibility and legitimacy from the South, but also wish to take the next step. That is, to be informed and reified by the South. This is no mean feat and climate change, as it would first appear, seemed almost purpose-built to traverse the globe in its unifying depiction of earthly experience. Only an invasion from outer-space could be more uniting, or so it was initially thought. But even with invasions, ultimately decisions of environmental protection, resource access, distribution and retribution surface once again. But for all these reasons and many others, human-induced climate change is seen amongst most environmentalists as *the* greatest issue which confronts humanity in the 21st century. Elsewhere we write:

> But climate change is more than just this. Not only does it bypass borders built by nation-states, the very concept decimates and invades collective identities forged by history, class, gender, race and caste. It redraws a map of the earth, at least a rhetorical map, as a single space occupied by all inhabitants, and casts them within a shadow of a global enemy – climate – something which cannot be seen or touched by most, but something which can only be interpreted and understood by a scientific and economic elite. The mapmakers, however, are not globally representative. The climate issue is largely a Western European initiative. It constructs geopolitics in a Western, realist sense, with its inherent notions of a polity occupied by nation-states acting rationally in a global anarchic system. (Doyle and Chaturvedi 2010: 520)

The 'climate project' has been criticized by some due to the way in which it has been utilized by the North, using science, to discipline the unruly

South for its moral recklessness in pursuing carbon-based economic solutions to poverty. As Simon Dalby has pointed out so insightfully, securitizing ecological issues helps provide justification for 'the global managerialist ambitions of some northern planners' (Dalby 1999: 296).

Climate change, then, in the first instance, exists at an elite, western level. Its very construction (indeed its very existence) is rooted in the global North. It has its origins in the scientific, anti-technocratic discourses of post-industrialist and post-materialist forms of environmentalism. Still today, the language of post-industrialism usually frames the climate campaigns environmentalists in the North. FoE Europe's website in 2010 reads:

> Climate change is the single biggest environmental threat facing our planet. Burning coal, oil and gas pollutes the atmosphere with greenhouse gases that cause the planet to heat up. According to the latest findings of UN climate experts (IPCC), temperatures could rise by up to 6.4°C before the end of this century. We must stay well below a dangerous temperature rise of 2°C. To achieve this we have to fundamentally change the way how we produce and consume energy ... To prevent climate change we need an energy revolution starting with an increase in energy conservation and energy efficiency, increasing the share of renewable energy sources ... Changing the way that we produce and consume energy can also create millions of new green jobs in sectors such as energy efficiency, energy saving and renewable energy. (FoE Europe 2010)

Much of this dominant trope relates to creating alternative forms of energy which do not necessarily undermine the projects of industrialization, modernity, technocracy, or capitalism, but replace these versions of Western-inspired, science-led *progress* with a version which is more *environmentally friendly* and more *carbon-neutral* (but still western-inspired and controlled).

The underlying answer to global climate change – the scientific and ecological modernist community tells us – is to *believe* the climate science. The problem is constructed in a manner that most of the Earth's people are ignorant, and need to be educated by the scientific elite into 'right thinking'. For example, in the North, Friends of the Earth (FoE) groups seek to educate as a part of '*campagnes de sensibilisation*' (Doherty and Doyle 2014), as a means of correcting *wrong thinking* (seeking to bring the wrong thinkers *into* their transient knowledge community). The contention is that once 'good science' has been understood by the masses (particularly those in the *non-integrating* peripheries of the global South), then the problem can be solved.

Of course, the other issue here is that 'good science' is constructed and sold in a classically modernist fashion, insisting that it is totally objective, and exclusively built on the external, empirical 'facts' provided to scientists, without any forms of social constructivism 'polluting the truth'. This widely made, closely held assertion by pro-human-induced climate scientists was largely the reason why it was so easy for climate deniers to attack their position during 'climategate' (Revkin 2009): the scandals that questioned the veracity and trust-worthiness of scientists working in universities and research institutions in the UK and the US (specific mention was made of the National Centre for Atmospheric Research in the US, and University of East Anglia in the UK).

By hacking into computer servers, a month before the Copenhagen climate summit, climate skeptics were able to argue that climate scientists 'conspired to overstate the case for a human influence on climate change' (Revkin 2009). Revkin reported the following in the *New York Times* (20 November 2009):

> The e-mail messages, attributed to prominent American and British climate researchers, include discussions of scientific data and whether it should be released, exchanges about how best to combat the arguments of skeptics, and casual comments – in some cases derisive – about specific people known for their skeptical views. Drafts of scientific papers and a photo collage that portrays climate skeptics on an ice floe were also among the hacked data, some of which dates back 13 years ... In one e-mail exchange, a scientist writes of using a statistical 'trick' in a chart illustrating a recent sharp warming trend.

Another scandal over the objectivity of 'climate science' emerged in relation to the IPCC's factually incorrect reports over a claim it made in 2007 that Himalayan glaciers would melt away by 2035 (Carrington 2010, 20 January). In early 2010, the then Indian environment minister, Jairam Ramesh, argued that although there were no doubts that the glaciers were receding, the claim was 'not based on an iota of evidence'. In a public relations statement put out by the IPCC after the event, it was admitted that the claim was false; that it had not been peer-reviewed; and that the IPCC had injudiciously relied on an interview report by environmental group, World Wide Fund for Nature (WWF).

Climate science, like any science, is constructed within epistemic communities. It is as much a product of human interaction as it is about essential properties of nature, derived through empirical observation. It is the very denial of the social construction phase which makes it, at times, so vulnerable. Science is made in communities, it is not perfect, it is incomplete, and it is a political process. Continuing to deny and occasionally hide these characteristics actually weakens the pro-climate case, rather than strengthening it. The work of Thomas Kuhn, on *The Structure of Scientific Revolutions*, is extremely apposite here. Written over 40 years ago, Kuhn's description of 'normal science' explains the two controversial examples above extremely well. He writes:

> Normal science, the activity in which most scientists invariably spend almost all of their time, is predicated on the assumption that the scientific community knows what the world is like. Much of the success of the enterprise derives from the community's willingness to defend that assumption, if necessary at considerable cost. Normal science, for example, often suppresses fundamental novelties because they are necessarily subversive of its basic commitments. ... the study of paradigms, including many that are far more specialised than those named illustratively above, is what mainly prepares the student for membership in the particular scientific community with which he will later practice. Because he there joins men who learned the bases of their field from the same concrete models, his subsequent practice will seldom evoke overt disagreement over fundamentals. (Kuhn 1970: 5)

Demeritt takes Kuhn's line of argumentation further, updating it and placing it firmly within current debates about climate change. Demeritt (2001: 307–337) writes on the problems of what he terms 'upstream' science; the creation of epistemic climate and global warming communities, and how members within these communities influence each other and their science. Demeritt aspires to overcome the epistemic and institutional distinctions normally applied between science and politics. He critiques the 'unreflexive', 'authoritative' understanding of science. In this article, he highlights the role of science presented as the apolitical supplier of neutral objective facts that informs the work of policymakers. In fact, the political needs of policymakers are obviously framing how science is constructed, and how the science is constructed, in turn, frames the responses of scientists. The 'objectivity claim', and the division of labor between scientists and policymakers, mask the fact that science has its own cultural practices and power relations, and politics has shared the practices and culture. Political preconceptions about how to manage climate change are at the root of the culture. The political insistence on demanding certainty from science (from scientists, politicians and the 'general public') is at the root of this culture, and this leads politicians to be absolved from the moral and political obligations and responsibilities underlying climate change. In addition, scientists are tacitly aware of the policy preferences of policy-makers and the need to frame scientific enquiry in a way that produces 'practical' outcomes. There are processes of 'mutual accommodation' and Demeritt argues that it is almost impossible to disentangle scientists' choices over questions and processes derived from their context. Extremely complex knowledge is produced and communicated to policymakers, who regard the knowledge with more confidence than do the scientists themselves. Demeritt contends that human societies must do away with the expectation that science can act as a 'uniquely privileged vehicle to Truth' (ibid.: 329). Instead, work needs to be embarked upon which will actually increase public understanding and trust in science (and scientists), based on the recognition of it being a social practice. He contends:

> Science does not offer the final word, and public authority should not be based on the myth that it does, because such an understanding of science ignores the ongoing process of organized skepticism that is, in fact, the secret of its epistemic success. (ibid.: 329)

Like Kuhn, Demeritt is a social constructivist, but – like the writers of this book – he does not think that social constructionist critiques of science are an argument for ignoring or rejecting climate science. Rather it is an argument for understanding science better. He rejects the claims of those from both the left and the right who use social constructivism in an instrumental way to critique science in support of their own political purposes. He takes as his theoretical base what he calls 'heterogeneous construction, which does not deny the ontological existence of the world, only that its apparent reality is never pregiven' (Demeritt 2001: 311). Some of these dominant constructions of the science of global warming are also based on other myths, judgments, underlying assumptions and experiential practices. He lists some of these as follows:

- Anthropogenic climatic change is a global-scale, environmental (as opposed to political or economic) problem.
- It is caused by the universal physical properties of GHS (as opposed to underlying political structures or moral failings).

- These objective entities have universal meanings that can be discovered scientifically by experts.
- The best way to understand global warming scientifically is to model it mathematically.
- An important objective of climate science should be the construction of more complex, comprehensive and physically reductionist models
- Model simulations provide the basis for future climate predictions.
- Rational policy is (or should be) founded on GCM projections about the regional-scale impacts of climate change.
- The regional scale is the most meaningful one for policy making
- Model parameterizations adequately simulate the climate system variability, or soon will.
- Modelers should focus first on (what they perceive to be) the most likely outcomes, as opposed to the most extreme.
- Experts are best placed to decide the legitimacy and credibility of these practices.

Demeritt concludes that most of these assumptions are not formal or institutionalized and are, in fact, only negotiated by small groups of scientists. In this way, a 'socially contingent form of scientific knowledge is being shaped by an emergent international policy regime' that, in turn, is 'being constructed and legitimated by this same body of scientific knowledge' (ibid.: 328).

This process, marked out in general terms by Kuhn and in the global-warming specifics of Demeritt, could largely account, therefore, for the controversies which we have listed above relating to the questionable pro-climate change email practices which took place at the National Centre for Atmospheric Research in the US and the University of East Anglia in the UK, as well as the IPCC scandal over the Himalayan glaciers. It is not usually a case, therefore, that pro-climate change scientists are deliberately and instrumentally cheating, or actively striving to 'fudge the data' (although this sometimes will happen). More often than not, by being positioned within the now dominant paradigm advocating the realities of human induced climate change (HICC), it becomes more of a structural issue. There can be no doubt that the weight of empirical evidence now supports their case. The paradigm, however, becomes an edifice, built as much upon empirical and experiential knowledge as upon communities, networks and institutions of like-minded scholars – with each new brick carefully chosen, its foundations must be protected by those that threaten it from outside its paradigmatic borders.

Unfortunately, in a geopolitical sense, these communities of climate scientists are almost exclusively based in (or at the very least, controlled by the minority world). Kandlikar and Sagar write at length of a South-North divide in climate change research in India (Kandlikar and Sagar 1999: 119–138).

At an international level, there was no emergence of North-South research, particularly in the formative, agenda setting years. The supranational research initiatives did not include substantive participation from scientists in Less Industrialized Countries (LICs). Funding usually supports scientists from their own affluent-world countries of origin, rather than enabling the work of LIC researchers. Southern researchers are connected to fellow majority world researchers only though pre-established Northern networks, and 'there is no exclusively Southern network whose focus on issues is not influenced by Northern interests', nor are there adequate attempts to coordinate or promote collaboration (ibid.: 132).

These structural problems lead to the effective silencing of southern scientists. Kandlikar and Sagar describe these dynamics as they are played out in the IPCC. The IPCC reviews and synthesizes nationally based studies. LICs necessarily are under-represented here. The IPCC is headquartered in the North, and networks are formed which physically exclude representation of southern scientists. Despite electronic communications, place remains important in network construction, and as a consequence, southern concerns are excluded from the IPCC agenda. This under-representation also tends to consolidate the dominance of certain disciplinary approaches of northern countries.

Indigenous 'Science', knowledge systems and climate change

Despite the early Northern-centrism of climate science, as the 21st century rolls well towards the middle of its second decade, climate change has increasingly established itself on southern agendas. It can no longer be understood in exclusively northern-centric scientific terms. But, as will be demonstrated throughout this book, in gaining acceptance in the South, climate change advocates have had to take on new cloaks in order to relate the issue to southern audiences. Much of this newer, more south-palatable ideology is dressed up not in the language of ecological modernization, but green post-colonialism (see Chapter 7). Sony Pellissery writes (2011):

> Climate science has generated deep societal divides in post-colonial countries such as India. These divides are 'constructing' the climate science in important ways. One of the most important divides comes from the fall out of the ideological positions created by

> post-colonialism. For centuries, religious and cultural systems that held the subcontinent India together were based on the principles of sacredness of nature, and the interdependent relationship that human beings enjoyed with nature. Indigenous science was built on this principle. This also celebrated diversity. Modernity and its science brought 'instrumentalism' into this relationship. Along with this, there was rejection of indigenous science, and thus the attempt for homogenization of human communities with the help of science into 'one nation'. The modern science was the accepted vehicle to remove poverty and to achieve development, and thus 'nation-building'. (2011)

Thus, more recently discourses have emerged which accommodate this neo-colonialism and, as a consequence, more southern-friendly forms of anthropogenic climate change rhetoric have emerged: 'The division of livelihood emission vs. luxury emission has emerged as an important contestation' (Pellissery 2011). With climate change gaining a foothold in the South, there have recently been strong attempts to also connect climate narratives with more indigenous knowledge systems and non-scientific knowledge systems to gain further credibility and approbation.

The other huge attraction of the climate change story is its close thematic associations with other ancient, conservative and archetypal narratives of mythological proportions. Most obviously, the story of the Christian Biblical Flood rings more than a few bells here. In this vein, global climate change is a cataclysmic event waiting to happen, with humanity to go the way of the wicked, untruthful and carbon-unfriendly (Doyle and Chaturvedi 2010). In this manifestation of the flood story, climate change encapsulates the concept of melting ice-caps, rises in sea levels and the flooding of small islands and coastal areas. God, or Gaia, is punishing us all for bad behavior. The North has the keys to the Ark; and, no doubt, are the chosen peoples to survive the first tsunamis. But climate change is even more compelling than Noah's flood, for it also incorporates the archetypal story of the great fire. For the flood waters will finally recede after the ice caps are long gone, and the Earth's temperature will continue to soar until the river systems dry up, destroying all agriculture by stripping the land of its soil, and entire cities will dry up as the heavens can no longer provide the very life source itself: water. And the wicked shall burn in hell, and the Day of Judgment will come.

Of course, the flood myth is far older than the Christian Bible. The Sumerian myth of 'Gilgamesh' predates it by at least a thousand years.

William Burroughs discusses the enduring fascination with the myth despite any lack of empirical evidence which confirms it as an event in 'real' time.

> The available climactic records do not contain evidence of some global cataclysm happening between 15 and 5 kya when most of the rise in the oceans occurred. There is nothing in the many climatic records to support various theories of sudden huge rises in global sea level associated with cosmic catastrophes or sudden shifts in the earth's axis or crust. Nonetheless, the widespread Flood Myth appears in many fables from prehistory. (Burroughs 2005)

But to limit these stories of cataclysm, the flood, the fire, the new dawn etc. to pre-Christianity and Christianity would be short-sighted. These are archetypal stories which cross many different cultures and societies. Obviously these stories have their equivalents in the other Abrahamic religion, Islam, perhaps also informed by 'Gilgamesh'. But they also ring with resonance in the Dharmic faiths of Hinduism and Buddhism. Anne Birrell writes of the power of the flood myth within the Chinese societal context:

> The most enduring and widespread of the catastrophe myths world-wide is the flood myth. In classical China the myth is told in four stories. The myth of the rebellious worker-god Common Work (Gong Gong) relates how he stirred the waters of the whole world so that they crashed against the barrier of the sky and threatened the world with chaos. The flood myth ... In this version the god Common Work plays the role of the marplot, one who seeks to destroy the design of the cosmos. In this respect, it is linked to the myth which tells how Common Work challenged the supreme sky god, Fond Care (Zhuan Xu), and in his fury butted against the world mountain that held up the sky. (Birrell 2000)

Now these great shared tales, which pop up again and again through the mists of human history, have resurfaced again, mixing with the religion of western science, and its priests – the scientists – have taken the mantle of the grand narrative's most enthusiastic seanachies. It is because the flood and the fire are stories which are bound into the very marrow of human existence that the climate change story enjoys such universal appeal, quite apart from whether the scientific data is 'real' or not.

Due to its wide cultural reach, its appeals to science (originally western and now more indigenous knowledge systems), and its conservative core (it challenges little regarding 'the order of things') the climate story is also unusually well-placed to serve a neo-liberal economic agenda, as shall be argued in Chapter 4. Environmental issues which emerged within the first flush of modern environmentalism demanded concepts such as limits-to-growth. Environmental issues were portrayed as a zero-sum game, with trade-offs demanded of business and development interests in order to achieve environmental goals. During the mid to late 1980s, ecological modernization and sustainable development discourses began to replace the finite growth, win-loss hypothesis, and we discovered that science, business and environmental interests could forge together, hand-in-hand, with voices advocating 'win-win-win' scenarios to become increasingly resonant across the globe.

3
Terrorizing Climate Territories and Marginalized Geographies of the Post-Political

Introduction

In academic and popular discourses alike, 'climate change' is often framed as a 'global challenge'; a threat beyond borders. This allegedly 'global' character of global warming (often taken as a defining feature of climate politics) portrays climate change as the paradigmatic global environmental problem. Its presumed globality links the climate issue to a broader discussion within international relations and critical geopolitics about the contemporary role of territory and political boundaries. The flows of people, capital and carbon across boundaries are perceived as indicators of a post-Westphalia world, stipulated and stereotyped as a deterritorialized and borderless political space. Climate change is thus contrasted in this discourse with a spatiality of global politics which is constructed as territorial, the parcelling up of the world into discrete political units. It is further approached and analyzed by some in terms of 'trans-national security threats', based on the geopolitical premise that 'predicted climate change impacts are also likely to strengthen or help revive sub-state networks that have traditionally responded to environmental change and pressure via violence, crime, smuggling, banditry, trafficking, terrorism, and other such activities' (Jasparro and Taylor 2008: 232).

Despite the overwhelming natural-science evidence in favor of a de-territorializing nature of climate change, as graphically revealed through various assessment reports of the Intergovernmental Panel on Climate Change (IPCC 2007), emerging geopolitical as well as geoeconomic discourses on climate change tend to (re)territorialize a whole gamut of issues at stake. Climate change geopolitical discourse is the discourse deployed by the 'winners' (and not losers) of corporate globalization. It is about controlling the contestation arising out of longstanding resistance against environmental degradation in many parts of global South.

In the body of this chapter, the current day cases of India and China are described, as these two so-called 'planetary powers' wrestle with the implications of a northern-centered climate map being superimposed, palimpsest-like, over the top of their own particular stories and territories of environment and development. Within the climate change and carbon-trading rhetoric – crossing national and cultural borders at will – the Indian and Chinese governments juggle their own domestic development needs – written large in their post-colonial roots – against their determination to deal with a global, largely post-industrial and post-political agenda (this supposed Global Soul), imposed on them by the North.

Of course, in the early stages of this book, it is crucial to reify a point made in the previous chapter: there are many *environmentalisms* (Doherty and Doyle 2006), both climate-centric and non-climate-centric, different green discourses which shape and configure conceptual and discursive maps. In short, there are key differences in constructions of environmentalisms in both North and South, outlining very different environmentally-determined realities in the daily lives of the many, versus those of an affluent minority. Through minority world lenses of post-materialism and post-industrialism, environmentalism challenges the excesses of the industrialist project; the rights of corporations to pollute and degrade; and the dwindling of the earth's resources as they are fed into the advanced industrial machines. Advocates of these positions (first emerging in the green movements of Western Europe) argue that advanced industrialism, championed by both the market systems of latter-day capitalism and the state-centered models of socialism, has pushed the earth, its habitats, and all its species (including people) to the brink of extinction. This industrial developmental paradigm has promoted economic growth at all costs. Initially this pursuit of growth, as discussed in the previous chapter, was rooted deeply in the Enlightenment project of the scientific and industrial revolutions, and the pursuit of progress. Hence the environment (and nature) was presented as an eternal cornucopia, where resources were unlimited.

In more recent times, industrialism has been globalised and homogenized. Now there is widespread and partial acceptance of natural constraints to growth – or a finite carrying capacity – but as discussed in the previous chapter, the Enlightenment project continues, as it now advocates increased growth through improvements to environmental efficiency and management, the promotion of the global 'free market', and the advocacy of homogenous 'democratic', pluralist systems. Most recently this has been pursued under the key terms *sustainable development* and *ecological modernization.*

Importantly, within these newer, adapted environmental discourses, there is a strong emphasis on the rights of 'future generations'. Indeed these rights are one of the defining features of sustainable development as it was first enshrined in the Brundtland documentation in 1987 (World Commission on Environment and Development 1987) as *inter-generational* equity. This fascination in 'future generations' is rooted firmly within post-materialism: the North imagines that its global citizen-consumers have achieved their basic material needs and, in neo-Maslowian style (Maslow 1943), can now pursue 'higher order' goals set for the citizens of the future.

But, in poorer and dominantly rural parts of the global South, movements for environmental justice and security are motivated by basic issues of survival for those who are already living, which are often the result of extreme environmental degradation and hundreds of years of colonial exploitation (Doyle 2008), rather than industrialization. So as an all-encompassing theory capable of explaining a global situational phenomenon, the post-material, post-industrialist and, to a large extent, *post-political* theses are found wanting.

Using post-colonialism as the narrative frame, green concerns are cast in the light of the colonizer versus the colonized; the dichotomous world of affluence and poverty. There are some obvious cross-overs with the previous post-industrialist thesis, recognizing structuralist lines between the haves and the have-nots. In different parts of the world, these frames, or story lines – and combinations of them and others – are used more often to explain the causes and effects of environmental issues and problems. In the global South, the frame of post-colonialism usually dominates.

Within the Northern, post-materialist, post-industrialist and post-political frames, it is often assumed that carbon-trading (discussed at length by us in the next chapter), built upon models accurately depicting the *carbon-footprint* of different nations, will be adequate to combat emissions. But within the post-colonial frame, carbon-footprints are largely meaningless, as they fail to adequately account for the past; they fail to adequately comprehend, and deal with, the concept of *environmental debt,* incurred through hundreds of years of Northern exploitation. Obviously there are some important exceptions to this rule: there have been some attempts to accommodate for the deeds of the past in the carbon-footprint literature of the North (Bayer et al. 2008). But on the whole, these attempts remain as subservient positions. Instead, the image of a 'divided planet' in terms of rich and poor, or 'Eurocentric planet', is a needed correction to the concept espoused by Ward and Dubos' *Only One Earth*, commissioned for the UNCED Conference in 1972 (Ward and Dubos 1987). Doug Torgerson goes further when he argues that the divisions of the planet bear the 'unmistakeable mark of the legacy of colonialisation' (Torgerson 2006).

In the next part of this chapter we argue that climate change is often used by the North to territorialize the world in a manner which dissolves and absolves differences between the affluent and less-affluent worlds; between the North and South in *post-political* terms.

The post-political, environment and 'cosmopolitan' climate change

The environment-welfare nexus is being approached and understood differently in both the global North and the global South (Catney and Doyle 2011; Mummery 2012). In the case of the former, environmental issues have been usually construed in a post-political manner, most particularly through post-materialist and/or post-industrialist lenses. Post-materialists largely see environmental welfare (through the rhetoric of sustainable development and ecological modernization) as something largely separate from humans (for the welfare of the 'rest of nature'). This sometimes manifests itself in discussion of the welfare of 'other species' or in arguments that insist that once the 'welfare of the planet as a whole' is pursued, then the welfare of humans will necessarily follow. When welfare is considered in the human realm, it is often viewed as something concerning future generations.

A key problem with this northern-centric, post-political (and often apolitical worldview) is that it conflates all categories of people into one. Difference is hereby dissolved. This notion is further supported by current arguments espoused by 'liberal cosmopolitans' (Osamu 2005), a trend particularly powerful in many parts of Western Europe and North America. In these arguments, by arguing in a post-political fashion, it is imagined the someone in the North who, for example, drives a Toyota 'Pious' (sic) in the UK for 'climate saving' reasons, is – in some way – alleviating the suffering of poor fishing communities on the Somalian coastline, in the South. The reality, of course, is very different. The politics of place and difference are still important and, however well-intentioned, the post-political floating (sustainability conscious) cosmopolitans of the global North only serve to further disempower their human 'compatriots' firmly rooted in the South, by denying the essential and elemental differences between these two largely incomparable worlds.

The current idea of the post-political 'future citizen' has become a central figure in environment and climate welfare debates, yet has been largely uncontested in environmental practice or thought and enforced differences remain intact. However, in the global South, the day of reckoning already exists. The metaphoric flood is in the past, not in a climate-changing future. In the South, crucial green welfare issues are almost always purely perceived in an anthropocentric manner, and most of these intersect with basic human rights: the right to have a healthy water source; the right to shelter; the right to food sovereignty; and the right to energy security. We argue that post-materialist and post-industrialist discourses do not recognize the colonialist realities of the global South, where people wrestle with massive environmental debts incurred upon them by centuries of exploitation by the North (the past and the present), rather than trading in sustainable climate footprints (the present and future). Our core argument is that there remain serious trade-offs between the politics of the now and the politics of the future, insoluble to the rhetoric of 'post-political' environmental discourses. Just because one looks to the future, and the rights of future citizens, does not mean that the people of the present are uniformly better off. Certainly, the converse is usually true (Catney and Doyle 2011).

Let us now explore the concept of 'post-political environmental welfare'. We start by outlining conventional accounts of sustainable development, pointing to the importance attached (at least in theory) to intergenerational justice considerations. We then turn to examine the 'post-political' critique of broad consensuses, such as that which exists over the necessity to place the consideration of the welfare of 'future citizens' over that of the 'present citizen' of the global South. We draw attention to the strategies and tactics of depoliticization deployed by global North nations, international institutions and NGOs to promote a consensus around which (post-political) sustainable development policies can be formulated and implemented. Having established the theoretical underpinnings in the second part of the paper, we interpret how key differences in the very manner in which the climate-welfare nexus is experienced and understood in both the global North and the global South are managed in favor of the former over the latter. We show in the final section of the chapter that post-political notions of environmental politics are nicely supported by liberal forms of cosmopolitanism, in trading off the rights and welfare of the majority world (living in the present) with those of an imagined 'green-white future' largely residing in the minority North.

Sustainable development, climate change and the post-politics of the future

The concept of sustainable development was popularized by the 1987 Brundtland Report (formally known as the World Commission on the Environment and Development) and endorsed by political leaders from across the globe at the Rio Earth Summit in 1992. Over the 1990s, the concept of sustainable development became embedded in the language of policy makers and academics to the point where it has been described as a new meta-narrative (Meadowcroft 2000: 370) or a 'neo-renaissance idea' (O'Riordan and Voisey 1997: 4). As Meadowcroft (2000) has observed, at a time when grand narratives and modernist projects to refashion man and society have gone out of fashion, a new 'meta-narrative' – one possibly even more ambitious than preceding projects – to guide future societal trajectories has arisen: sustainable development. In this sense, the 'future' becomes an object for action, requiring the resurfacing of planning as a key mechanism for guiding public policy, private business and individual conduct (Raco et al. 2008: 2566).

Sustainable development (SD) is often criticized for its ambiguity. This ambiguity has given rise to a panoply of discourses that interpret sustainability in a variety of ways within and across nations (see Lafferty and Meadowcroft 2000). The dominant interpretation of sustainable development comes in the form of the theory of ecological modernization (EM) discussed in the previous chapter. SD can be understood as an attempt to resolve the traditional tension in environmental politics between striving for economic growth and protecting the environment. Instead of seeking to replace capitalism with some other alternative system of socio-economic organization, SD suggests that governments, corporations and civil society can seek to promote economic growth but that they must take greater responsibility for protecting the global environment from further damage. These arguments are taken further in the next chapter, when we directly address the relationship between discourse of the market and climate change. Suffice to say here that advocates of SD argue that it is possible to decouple economic growth from environmental harm and that the application of new technologies and the redesign of institutions can reduce or better manage the amount of raw material throughput, energy use and waste generation that modern societies produce. The Brundtland Report (World Commission on Environment and Development 1987) supported this interpretation of sustainable development by arguing that continued economic growth could support environmental protection, as well as promoting social development (Doyle 1998).

EM and SD have been key components of an influential discourse in the global North over the past two decades, utilized in UK environment policy. Both Revell (2005) and Barry and Paterson (2004) argue that the UK has embraced a weak version of EM and SD, with politicians from both Conservative and Labour parties deploying the language of SD in their election manifestos and speeches on the environment (see Salih 2009). At a time of increased ecological awareness, the 'win-win' (Revell 2005) philosophy of EM enables politicians to offer policy solutions to these threats that also contribute to economic growth. A reduction in economic growth and consumption rates, it is often argued, would limit the degree to which the state can make public welfare provision. Already, welfare states in the global North are confronting a number of post-industrial challenges such as low levels of economic growth, ageing populations and changing family structures (Pierson 2001). States are hence required to push further for economic growth (and, indirectly, social welfare), potentially at the expense of the environment. For example, while the UK government has made considerable progress in terms of ecological modernization over past two decades, the economic imperative has intervened to push anti-environmental agendas, such as the growth of air travel to secure inward investment and promote growth (Barry and Paterson 2004).

It is our claim that 'sustainable development' is part of a broader process of 'post-politicalization'. Before examining what is considered to be 'post-political' about the present condition, it is first necessary to consider 'the political'. 'Politics' is often used as a term to denote the institutional sphere of the state and the organized competition for votes and governing that takes place within its boundaries. For Gamble (2000: 1), by contrast, the political opens up a realm where human societies can 'seek answers to fundamental questions of politics – who we are, what we should get, how we should live.' It is the space where there is a 'constant clash of interests, ideologies and values, generating rival parties and movements, alternative principles of social and economic order, and competition to realize them' (Gamble 2000: 1). This conception of the political emphasizes the importance of the clash of alternative visions of future societal trajectories. The development of such a political space requires the 'public encounter of heterogeneous groups and individuals' with often radically different perspectives on what future direction society should take (Swyngedouw 2008: 4). As Marchart

(2007: 42) perceptively observes, distinguishing between 'politics' as a noun and 'the political' as an adjective is by no means unproblematic but it is analytically useful in *de-territorializing* our understanding of the political, uncoupling the tight bond between political phenomena and institutional configurations.

As with all uses of the prefix 'post', misunderstandings can arise. The 'post-political' is not used to connote the end of politics *per se*; the formal institutions of government (particularly in the global North) remain important sites of power, political parties continue to be active and elections are held at regular intervals. Post-political theorists (Valentine 2005) argue that the space of the political is contracting in the face of the hegemonic grip that neo-liberal ideas have over public affairs (Catney and Doyle 2011b). The political realm is increasingly limited to managerial concerns over 'what works' (as if these decisions were not themselves 'political') than with the clash of competing alternative principles of social and economic order.

Post-political theorists claim that 'consensus' is promoted as a means of closing down debate about larger issues relating to political economy or existing societal power relations (Paddison 2009: 5). As Swyngedouw (2007: 24) observes, post-politics is 'about the administration of social or ecological matters, and they remain of course fully within the realm of the possible, of existing socioeconomic relations.' Authors such as Žižek (1999a) Ranciere (2007), Dikec (2005), Swyngedouw (2008; 2009) and others, argue that a key factor behind the rise of a post-political order is the accelerated 'policing' of politics and policy processes by 'bureaucrats' and 'experts' who seek to naturalize particular governance arrangements and privilege certain ideas and interests. Governing becomes a matter of reducing disagreement and promoting consensus over the parameters of discussion so that politics becomes, as Valentine (2005: 55) argues, 'a matter of maintaining a minimum level of cybernetic equilibrium within circumstances which it does not authorize and disagreement is reduced to the status of a practical problem in search of a solution.'

At the heart of post-political governance is the need to displace dissent and manufacture consent to prevent the politicization of policy (Swyngedouw 2008: 10). It is argued that such an approach promotes 'good governance' – a term readily recognizable in the discourse of institutions such as the World Bank (Harrison 2005) – because important issues are not drawn into political disputes that reduce the scope for reasoned reflection on the optimal policy solution. For critics of post-politics, participatory mechanisms are not a supplement to democracy, enhancing the opportunities for outside voices to be heard in the policy process, or a way of improving the rationality of policy processes, but are merely a way of manufacturing consensus and thereby limiting dissent.

Swyngedouw (2007: 27) argues that the discourse of sustainability seeks to evacuate the potential for radically alternative socioeconomic and socio-environmental orders by placing limits on 'the possible', by marginalizing or seeking to silence radical antagonisms. In short, Swyngedouw (ibid.: 27) views climate change as representing the negation of the political and the promotion of post-politics. For Swyngedouw (2007: 26–27) the construction of post-political environmental consensuses

> is one that is radically reactionary, one that forestalls the articulation of divergent, conflicting, and alternative trajectories of future socio-environmental possibilities and of human-human and human-nature articulations and assemblages. It holds on to a harmonious view of nature that can be recaptured while reproducing if not solidifying a liberal capitalist order for which there seems to be no alternative.

We argue that a particular global North (post-materialist, intergenerational) conception of sustainability has come to dominate conceptions of welfare and even the governance of the state in the global South, marginalizing more immediate welfare concerns in these states. Climate change *is* a key discursive site where this post-politics takes place; it is, on many occasions, a *depoliticizing* clutch of green ideas and actions.

Depoliticization

Within the concept of the post-political, depoliticization is a tactic intended to take 'the political' out of decision-making over difficult issues (see Flinders and Buller 2006). For the purposes of our analysis here, we conceive the 'post-political' as a concept that operates at the macro-level, with 'depoliticization' operating at the meso-level as a tactic that operates as the constitutive process of the post-political (ibid.). As Flinders and Buller observe, the concept of depoliticization is something of a misnomer as politics remains. Indeed, it is often politicians that decide what is 'political' and what is not – although it operates to a 'narrow interpretation of "the political" that largely focuses on institutions and individuals commonly associated with representative democracy' (Flinders and Buller 2006: 296).

It is a strategy that is mainly justified on the grounds of specialization, that is, that certain issues demand esoteric, specialist knowledge to ensure optimal decision-making (Flinders and Buller 2006: 300). Faced with complex and/or intractable policy problems such as climate change and other controversial environmental hazards, power is increasingly placed in the hands of technical-scientific experts who are then asked to offer recommendations for 'solving' (or at least mitigating) the crisis (Swyngedouw 2008: 10). Often this process is forced upon (global South) states by institutions such as the International Monetary Fund or World Bank, as a means of promoting 'better' development (Eggertsson and Eric Le Borgnesee 2007; see Harriss 2002; Ferguson, 1994 for a critique). Depoliticization is a tactic for taking issues outside the realms of politics, to keep them away from the sphere of political contestation.

Neopopulism

Paddison (2009: 3) tentatively asserts that a new style of neopopulist governance has emerged which fosters a range of strategies which seek to advance neoliberal policies, though principally consensus building, persuasion and even coercion (for example, outlawing forms of protest) are part of the politicians' repertoire. Swyngedouw (2009) more directly links the rise of neopopulism to the rise of post-political governance and the advancement of neoliberalism, arguing that politicians engaged in neopopulist strategies seek to invoke the rhetoric of external 'threats' against a unified 'people' in order to build a consensus through which alternative narratives are foreclosed. For example, globalization or climate change are projected as threats to the 'the people' or 'the city' which requires consensus-oriented political responses, which in turn relegate the importance of other questions that lay beyond the immediate resolution or amelioration of the 'threat' at hand (Paddison 2009; Swyngedouw 2009).

The unit under threat ('the people', 'the city', 'the nation', even 'the planet') is spoken of in unitary terms, as collective with a common cause and common purpose, in Schmittian terms as 'friends', while the threat is couched in terms of an 'enemy'. Rather than being seen as a politicizing process *pace* Schmitt, an inversion is claimed to take place that constructs an enemy and builds a consensus which prevents attention being given to other issues such as unequal societal power relations and (environment-social) injustice (Paddison 2009; Swyngedouw 2009). Importantly, institutionalized forms of participation are a crucial part of this approach:

> Neo-populist strategies do not just emphasize the unity of the people but are active in demonstrating that the people are part of the political process, hence the emphasis given to political participation. How, though, participation is performed – and what issues are debated – becomes constrained to the agenda needed to pursue economic objectives. (Paddison 2009: 8)

Similarly, Rosanvallon (2008: 254) observes that in depoliticization there is often *greater* involvement and participation of civil society in politics and policy than there was previously. The construction of participatory mechanisms, alongside discursive constructions of threats and commonality of purpose, are perceived by critics of 'post-politics' in manipulative terms, as a means of staving off issues that are beyond the acceptable terms of the consensus. As we shall show below, by examining concepts such as liberal cosmopolitanism (and sustainable development) through a post-political lens, we can observe how these ideas can be used to promote a sense of solidarity through the development of a 'global we' which is dominated by the rationalities, and serves the interests of, the minority global North.

The analytical value of the post-political thesis is in its sensitizing of the potentially exclusionary nature of contemporary environment/climate debates and governance processes, in particular in constructing certain (depoliticized) 'futures' which act to exclude alternative perspectives. The post-political offers a useful heuristic for understanding how the discourse of sustainable development can come to dominate and silence alternative future societal trajectories, in particular marginalizing the urgent welfare needs of peoples in the global South. We recognize that the 'post-politics' thesis is not without its problems, relating principally to the way in which it can downplay the extent of agency open to actors in challenging 'consensus politics' (Paddison 2009: 7). For example, new social movements have challenged 'consensus' in the past and have been critical in advancing social concerns such as the rights of women or promoting ecology, or have undermined governments and challenged economic orthodoxy (Offe 1985; Drache 2008). Making a binary distinction between political and post-political or politicized and depoliticized forms of governance risks the accusation of crude reductionism, of oversimplifying complex environmental debates or governance processes (Flinders and Buller 2006: 297). Furthermore, it is often unclear who or what, if anything, is directing the trajectory towards post-politics, although there are a number of potential sources, including: the rise of hegemonic neoliberalism; an increasingly apathetic and conservative citizenry that has lost confidence in the capacity of the political system to create change, thus placing limits on the scale of ambition on the part of politicians and bureaucrats; and/or the (un)intended consequence of state restructuring processes over the previous three decades.

Sustainable development, liberal cosmopolitanism and the citizen of tomorrow

As noted above, Sustainable Development suggests that we need to have greater regard for the welfare of future generations so that they can enjoy similar resources and opportunities to the ones we presently enjoy, but at the same time offer the prospect of advancing the living conditions of people living in the global South (Jacobs 1999). It is here that sustainable development is visibly bound up with a contested *'politics of time'* (Raco et al. 2008: 2655), or more precisely, a *'post-politics of the future'*. We argue that a divergence has opened up between the environmental values of the global North and those of nations in the global South. As will be discussed below, this divergence has resulted in contrasting interpretations of the proper timescales for sustainable development (also see Meadowcroft 2002), which brings within its scope debates about the nature of welfare and ideas of justice across time and space. Within this 'post-politics of the future', three figures become visible: the (largely forgotten) 'past citizen', the (condemned) 'present citizen' and the (post-political) 'citizen of the future'.

The 'past citizen' is perhaps the figure most obscured in the debates on climate change. We argue that there is the 'debt of the past' in terms of rapid population growth and the accumulated weight of past carbon-intensive activities. By conventionally establishing 1990 as the 'baseline' for measuring 'sustainability' we are essentially denying the debt of past activities by the global North by locking future patterns of 'sustainable development' within the parameters set for nations that successfully industrialized first. Few have argued for the 'baseline' to be established at 1960 levels, when the difference in development terms between the global North and South was less pronounced. The citizen of the present – particularly in the global South – is a condemned figure in that s/he is held to be equally responsible for correcting for the debt of the past (generally that of the global North) whilst paving the road to a (post-political) future. The 'future citizen' is often framed through the ideas of liberal cosmopolitanism: as post-political (or even apolitical). It is one that is held to be beyond nationality, class, and race.

However, despite the post-political nature of sustainability discourses, there remain serious trade-offs between the politics of the now and the politics of the future. These challenges prove insoluble to the rhetoric of 'post-political' environmental discourses. Just because one looks to the future, and the rights of future citizens, does not mean that the people of the present are uniformly better off. Certainly, the converse is usually true. It does not take much imagination to drift back to the early years of the modern environment movement and be forced to revisit eco-conservative Paul Erlich's work, *The Population Bomb* (1971). Erlich's 'life-boat' Earth was roundly criticized, not for placing limits on the Earth's carrying capacity, but for discussing 'eugenic' qualifications for whom and what should gain priority to access the survivalist boat. These early arguments were full of nascent racism and cartographic anxiety (Chaturvedi and Doyle 2010). Advocates of green post-political ideas included in the bastions of post-materialism and post-industrialism, ably supported by liberal cosmopolitans, imagine no differences between those in the lifeboat and those outside of it. These thinkers (they rarely have their feet in the clay of survival) construct a dangerous *universal We*. There are, in fact, profound costs to the citizens of the global South living in the present, in any green political equation which trades their present rights and welfare with those of an imagined post-political and globalized neo-liberal future. In effect, with reference to the earlier documentation on sustainable development, we must ask, 'Whose future are we sustaining?' A green one or a white one?

Post-materialism versus survival?

As noted above, sustainable development suggests that we need to have greater regard for how future generations can enjoy similar resources and opportunities to the ones we presently enjoy (see Dobson 1999). This idea of intergenerational justice is one more prevalent in the increasingly post-materialist environmental discourses of the global North than the South. Strongly premised on Maslow's (1954) 'hierarchy of needs', post-materialists such as Inglehart (1990) argue that having largely fulfilled its more basic needs of safety and security, parts of advanced industrial society are able to pursue the 'higher', more luxuriant causes of the world – such as love and a sense of belonging (or even the rights and welfare of future generations) – beyond the politics of present-day and material existence. As a result, such a value change sees the welfare of non-humans being given greater consideration (Hay 2004). Indeed this new ground in considering the rights and welfare of 'other nature' is quite inspirational and, indeed, at the forefront of radical green thought. But, despite recognizing the benefits to all inhabitants of the globe – both human and non-human – within a more holistic and inclusive view of the welfare of the Earth, this conceptual framework can also be quite negligent of welfare issues in the global South as it often lumps Homo sapiens together as one entity, further diffusing welfare differences between the global *haves* and the *have-nots*. As Miller (1995: 146) notes, the issue of equity for future generations and equity for present generations in the global South that lack access to the types of goods needed for development (or even survival) is one that is increasingly voiced in international debates on the environment: 'An adequate response to the global environment will therefore have to address issues of equity; it will also have to be based on an awareness of the variety of ways in which the environment is socially constructed in different parts of the world.'

While post-materialist values arguably dominate environmental thought in the global North, the global South still wrestles with the more basic needs of survival for those actually living on the planet, rather than those who might sometime in the future. As Doyle (2008: 311) has observed and as we will show in the next section it is through the policies of development institutions and international environmental NGOs that post-materialist approaches to environmental protection and welfare are imposed on the global South in place of (sustainable) development trajectories which are more firmly rooted in local political rationalities. This imposition of a post-materialist frame, we argue, has in some parts of the global South been achieved through the construction of 'consensual' institutional arrangements that favor voices firmly located within post-materialist-orientated discourses that promote 'intergenerational justice' and overlook immediate survival concerns of populations in these parts of the world. Through neo-populist rhetorical plays on 'persevering the future for the next generation', a consensus is formed which urges dramatic reductions in carbon emissions and forecloses considerations of a more fair and equitable distribution of global resources.

A post-politics of the future, and the politics of the now and the past

In Chapter 7, we look in detail at how non-state actors, particularly in the environmental movement, are engaging with climate change. But due to this constant harping on behalf of minority world by writers advocating the existence of an omniscient post-political realm, the barriers between state and non-state actors is increasingly blurred. As a consequence, we have to briefly touch upon non-state actors here and their engagement with climate change.

It is our contention here that 'the future' and issues of intergenerational justice have become central parts of the construction of an environmental (post-political) consensus in which the claims of the 'future citizen' of the North takes precedence over the current citizen of the global South. Through the employment of tactics such as active depoliticization, a post-political settlement is constructed that takes politics out of issues of global distributive justice such as over resources and welfare questions.

For post-political theorists like Žižek (1992) and Swyngedouw (2007), a starting point in any discussion over sustainability needs to address what it is that needs sustaining and why. While accepting hazards posed by environmental degradation and climate change, these theorists pose challenging questions on what is the Nature that we are sustaining. As Swyngedouw (2007: 19–20) has observed, 'imagining a benign and "sustainable" Nature avoids asking the politically sensitive but vital questions as to what kind of socioenvironmental arrangements we wish to produce, how these can be achieved, and what sorts of nature we wish to inhabit.' According to these theorists, sustainability is based on a fictional conception of nature which is fundamentally harmonious and which, through the application of various technical and managerial 'solutions', can be returned to a state of equilibrium (Swyngedouw 2007: 23). Seen in this light, sustainable development can be interpreted as a post-political phenomenon that seeks to preserve established socio-economic relations rather than challenge them.

On similar grounds, the notion of 'intergenerational justice' becomes a problematic concept. Beckman and Pasek's (2001) exploration of the relationship between justice and posterity found that the very notion of establishing a concept of justice across generations is elusive and difficult to establish. However, the notion of intergenerational justice has become increasing prominent in the increasingly post-materialist discourses of the global North. We do not wish to rehearse here the various constructions and critiques that have emerged in the area of intergenerational justice (see Dobson 1999; Attfield 2003: 96–125; Page 2007a, 2007b); rather our concern is how notions of intergenerational justice are part of a post-political consensus which seeks to close down debate over current development pathways, thereby reinforcing existing socio-economic relations, while promoting the necessity of a 'shared burden' between the global North and South in terms of carbon reduction.

At global environmental summits, the consensus has been demonstrated to be shallow. For example, in July 2008, the leaders of the G8 (Canada, France, Germany, Italy, Japan, Russia, the United Kingdom, and the United States, plus a representative from the European Union), met with the leaders of the eight emerging economies (Australia, Brazil, China, India, Indonesia, the Republic of Korea, Mexico and South Africa) to agree a 'shared vision' on how carbon emission reductions could be achieved. From this summit, a joint communiqué was issued which stated that:

> Climate change is one of the great global challenges of our time. Conscious of our leadership role in meeting such challenges, we, the leaders of the world's major economies, both developed and developing, commit to combat climate change in accordance with our common but differentiated responsibilities and respective capabilities and confront the interlinked challenges of sustainable development, including energy and food security, and human health.
>
> (BBC 2008)

The joint communiqué seemed to take seriously the long-term concerns of the global North (sustainable development) and the more immediate concerns of the global South (energy and food security, and human health). However, it also gave an indication of where key differences and cleavages within the consensus existed. The shallowness of the consensus was exposed one year on at the UN Climate Summit in Copenhagen in December 2009 when global South countries walked out of talks, angry at what they saw as wealthier nations seeking to sideline Kyoto Protocol targets for reducing carbon emissions. What this episode demonstrates is the elusive nature of consensuses over climate change and sustainable development. While a superficial level of agreement could be achieved that some form of action needs to be taken to secure a future for the planet, the consensus rapidly unraveled when more precision was required over how 'sustainability' was to be framed and what the timescales should be for action, and how the burden on achieving this should be achieved. We will be returning to some of these issues in the concluding chapter of this study.

The aforementioned climate change issue is an excellent – and most recent – example of this construction of universal We, a global Us. It is green post-politics in its purest form. As we have written elsewhere:

> A critical view of the predominant climate change discourse is that it takes much of the politics – the conflict – out of environmental

> resource issues, providing a polite filter between human action and human consequence; taking the direct and instrumental power relationships out of the equation. It is no longer people against people: the exploiters versus the exploited, or in this case, the polluters versus the polluted. Rather, although people are still the initiators, they are cast in a far more oblique light, often unwittingly setting off a calamitous, climactic punishment for all. A force of nature is, in the end, the nemesis, whereas the initiators, the environmental degraders, are in relative safety, at a convenient one step removed from the atrocities inflicted upon the many. Also, by constructing the concept of an environmental 'day of judgment' for all, all humans (all creation) are cast equally as victims; not differentiating between the perpetrators and fatalities.' (Doyle and Chaturvedi 2010: 533)

The climate change agenda is largely constructed by Northern actors and imposed on the South. The case study of Friends of the Earth International provided in Chapter 7 illustrates empirical evidence of profound dichotomies between affluent and less-affluent world groups within this global federation of environmentalists. In a questionnaire designed to measure which national groups were involved in which campaigns, it was revealed that the vast majority of FoE national groups involved in climate change activism were based in Europe (Doherty and Doyle 2011). FoE activists in Africa, Latin America and the Caribbean, and in the Asia-Pacific were far more likely to be involved in food sovereignty, or water security campaigns. Furthermore, the post-industrialist emphasis of climate change campaigns, supported and constructed by elite, western science is seen as another form of imperialism, this time wearing a green cloak, directly challenging the rights, development and welfare of those living in the South. Also, in true post-political fashion, amongst the Northern groups, there is much store put in 'education programs' to inform the people of 'correct thinking' in scientific, 'carbon-friendly' terms; whereas in the South, where political communities still *exist*, there is an understanding that subservient models of knowledge which abide in, and emerge from, these communities are more appropriate in providing welfare policies and prescriptions than those evolving from externally-imposed, elitist science. Again, this conflict between the fundaments of the political and the post-political – dressed up this time in the fabric of climate change – was one of the key reasons for FoE Ecuador leaving the transnational green organization (see Chapter 7).

Finally, when green NGOs in the South do engage with the climate change debates, they are far more likely to pursue models of carbon-debt and dialogues of reparation for the carbon-inspired sins-of-the-past. Furthermore, they do not accept the Northern climate agenda which prioritizes the importance of a depoliticized, non-regional, non-time specific, cosmopolitan system of measuring carbon footprints, which forms the base-line data for neo-liberal models of carbon-trading and accounting, explored in Chapter 4.

Green deterritorialization as advanced capitalism: the construction of a *Global Soul*

In their intriguing study of global social movements, Graeme Chesters and Ian Welsh (2006: 6) write of a global mobility central to the prevailing capitalist axiomatic, which opens up physical borders through the imperative of deregulation, defining 'globally extensive sets of rule-bound domains establishing the primacy of the prevailing capitalist axiomatic over local custom, traditions and rules' (ibid.). This is the process which, in the theoretical footsteps of Deleuze and Guattari (1987), they would refer to as deterritorialization. They write:

> Ultimately, the projected potentiality of this global institutional nexus is carefully honed through external relations and marketing divisions to present a positive immanence within the public sphere. This is a process of deterrietorialisation which seeks universal benefits of the prevailing axiomatic removing barriers to implementation, effectively rendering space a 'smooth' obstacle-free surface ... The potential emphasised is the win-win face of globalisation as freedom, prosperity, choice and affluence. (Chesters and Welsh 2006: 6)

In a biophysical sense, climate, like the wind, crosses boundaries, moving in, through, over and under the politics of nation-states. But climate change is more than just this. Not only does it bypass borders built by nation-states, the very concept decimates and invades collective identities forged by history, class, gender, race and caste. It redraws a map of the earth, at least a rhetorical map, as a single space occupied by all inhabitants, and casts them within a shadow of a global enemy – climate – something which cannot be seen or touched by most, but something which can only be interpreted and understood by a scientific and economic elite.

The mapmakers, however, are not globally representative. The climate issue, as mentioned, is largely a western European initiative. At its moral core, it is fundamentally conservative. It constructs geopolitics in a western, realist sense, with its inherent notions of a polity occupied by nation-states acting rationally in a global anarchic system. But its foothold in anarchy is even deeper than this, for it conceptualizes *the natural order* as anarchical as well. At the core of this system is a fear of chaos which will be unleashed upon the world if a centralized, moral (carbon-neutral) authority is not maintained. In this neo-Hobbesian sense, nature is at war and, in the context of climate change, the earth's people will fall into a sea of chaos if Westphalian understandings of order are not imposed. This conservative moral core is used in a manner that seeks to discipline the unruly South for its moral recklessness in pursuing carbon-based economic solutions to poverty. As Simon Dalby (1999: 296) has pointed out so insightfully, securitizing ecological issues helps provide justification for 'the global managerialist ambitions of some northern planners'.

Importantly, carbon trading and other post-political and neo-liberal (see next chapter) solutions can only succeed if the very nature of the liberal subject is challenged, and turned into a neo-liberal subject: the global citizen-consumer. The existence of one *grand narrative* of flood and fire aids this process of atomization. With the global polity understood as a 'pluralist' one, played out on a notional level playing field, each citizen is actually a consumer. All people are considered 'equal' under this model; they just need *listening to*, or entry into the market place. There is no longer a clear delineation between *us* and *them*; haves and have-nots; subjects and objects of environmental degradation.

This view of the Earth comprising a series of human, interchangeable, individualized parts fits in neatly with the market. All beings are seen as consumers and providers, all meaning systems of collective action which may generate opposition are done away with because: 'We are all just people, just global citizens'. Questions of gender, class, and race – it is imagined – melt away.

Concepts of *intra-generational equity* are dismissed and replaced with the post-material concept of *inter-generational equity*. The only differences deemed valid are those demarcating the living from those who *may live* sometime in the future. In this manner, humanity is made faceless, and the rights of the present are sacrificed to imagined apocalyptic futures.

In turn, it is proposed that the salvation of the earth will come from *within* individuals, not in the politics of communities. Nature and its peoples are further commodified within this framework. The principles of sequential use and climate change are built on a commitment to global values change and *'tweaking' the capacities of people*, whilst denying the existence of any power differentials between cultures and people. This is the construction of a *global soul*.

Furthermore, the unwritten text of this process is that there is something wrong with the values – the Soul of the South – of the *majority world*. Environmental problems, like climate, are seen as getting worse due to the incorrect value systems of the South, rather than from the pressures on local communities to remain competitive in the new globalised, free market-place. Of course, the dissolution of the concept of *the Other* has not occurred in the South: in fact, the boundaries have often solidified with the acknowledgment that globalisation has delivered disproportionate amounts of wealth and power to Northern and Southern elites (Doyle 2005). It is critical, therefore, to re-assert a clear demarcation, despite the best attempts to create this global soul, between the environmental realities of the less-affluent majority, from the more affluent minority.

Spatial framings of climate change: re-territorializing the deterritorial? the case of India and China

Of course, this building of a global soul – with its in-built denial of the existence of any legitimate opposition to its agenda – may spell the death of boundaries based on rich and poor, on gender, on race; but even its most ardent supporters must concede that it clearly does not eradicate differences between nation-states. In fact, in several ways, climate discourses re-affirm the primacy of nation-states. These discourses are significantly marked by what John Agnew (1994) has described as the 'territorial trap': imagining the world as a series of rational and spatially and politically distinct states. That there is a complex geographical-spatial politics to climate change, in which both corporate and individual actions play a central role is conveniently glossed over by the state-centric spatial imaginary (Barnett 2007).

As pointed out by Lovbrand and Stripple (2006), various deterritorial representations of the atmosphere and climate problem notwithstanding, international climate policy over the years has resulted in territorialization of the carbon cycle. The discursive-geopolitical transformation of global and deterritorial carbon cycles (and its concomitant science) into territorial 'national sinks', dictated and driven by the territorial framing of terrestrial carbon uptake, argue Lovbrand and Stripple, '... can only be understood with reference to the inter-governmental negotiations on climate change. Since the parties to the climate convention decided to adopt a net-accounting of national greenhouse gases, a whole new repertoire of accounting methods and techniques have developed to standardize the national reporting of changes in carbon pools embedded in vegetation and soils' (ibid.: 234).

India's official framing of climate change issues continues to oscillate between the 'scientific' imperatives of deterritorialized-global understandings of climate change and reterritorialization of climate space through geopolitical-geoeconomic reasonings. The tone and tenor of India's climate change geopolitical discourse, against the backdrop of what we would like to describe as the 'revolution of rising socio-economic expectations', is quite visible in the 2006 National Environmental Policy (MoEF 2006), which lists the following elements as central to India's response to global warming: adherence to the principle of common but differentiated responsibilities and respective capabilities; prioritization of the right to development; belief in equal per capita entitlements to all countries to global environmental resources; reliance on multilateral approaches; and participation in voluntary partnerships consistent with the UN Framework Convention on Climate Change (UNFCCC).

India has argued in international fora that responses to and action on climate change must be based on science and not treated as a 'post-modernist religion'. However, the then Indian Prime Minister, Manmohan Singh, was quick to point out in his 2007 address to the Indian Science Congress that, 'the science of climate change is still nascent and somewhat uncertain' and called upon the Indian scientists to further 'engage in exploring the links between the greenhouse emissions and climate change' (GOI 2007).

In response to anxieties expressed by some that India will become the third largest emitter by 2015 (and, together with the US, EU, China and Russia, it will account for two-thirds of the global greenhouse gases), and hence should commit to certain emission reduction targets, the Indian official discourse runs as follows: India, given its limited role in contributing to the problem thus far, coupled with its compelling developmental needs and the historical responsibility of the developed countries, cannot be expected to take on mitigation targets. Further cited in favor of this reasoning are the findings of modeling based on the Integrated Energy Policy, which

> demonstrates that in the worst case full-coal scenario by 2031–2032, India's per capita emissions will be 3.75 metric tons per capita, and in the best case scenario with full use of renewable, maximum use of nuclear, hydro, and natural gas, significant increases in coal efficiency, and a 50 per cent rise in fuel efficiency of motorized vehicles, per capita emissions will be 2.66 metric tons per capita (Rajamani 2009).

Given these premises, the argument then concludes: Given the substantial cost that such a reduction will give rise to, it is not worth the benefits even to the international climate effort. Instead, India would advocate equitable emissions entitlements to the atmosphere. The former Indian minister for Environment, Saifuddin Soz, is reported to have said at Kyoto, [p]er 'capita basis is the most important criteria for deciding the rights to environmental space. This is a direct measure of human welfare. Since the atmosphere is the common heritage of humankind, equity has to be the fundamental basis for its management' (ibid.).

The following excerpt taken from the text of the address delivered by the former Indian Prime Minister Mr Manmohan Singh on the release of India's Action Plan on Climate Change (NAPCC) on 30 June 2008 (GOI 2008) graphically shows the tension between the global/deterritorial and national/territorial logics deployed in response to the dilemma faced by a 'Rising' India with 300 million plus middle class; a class that symbolizes India's status as a rising Asian power on the one hand, and, at the same time, can be held as most responsible/accountable in terms of per capita emissions to the 'global' atmosphere.

> *Climate Change is a global challenge.* It can only be successfully overcome through a global, collaborative and cooperative effort. India is prepared to play its role as a responsible member of the international community and make its own contribution. We are already doing so in the multilateral negotiations taking place under the UN Framework Convention on Climate Change. The outcome that we are looking for must be effective. It must be fair and equitable. *Every citizen of this planet must have an equal share of the planetary atmospheric space.* Long term convergence of per capita emissions is, therefore, the only equitable basis for a global compact on climate change. In the meantime, I have already declared, as India's Prime Minister, that despite our developmental imperatives, our per capita GHG emissions will not exceed the per capita GHG emissions of the developed industrialized countries. This should be testimony enough, if one was needed, of the sincerity of purpose and sense of responsibility we bring to the global task on hand. (emphasis added)

The position taken by India at the G-8 Summit held in L'Aquila, Italy, in July 2009 generated a controversy over India signing the declaration of the Major Economic Forum (MEF) on energy and climate that was held alongside the summit. The critics point out that by endorsing the following declaration India has admitted a cap on its emissions, which would undermine both development efforts and the stand taken by India all along that it will not accept any legally binding limit on its emissions:

> We recognize the scientific view that the increase in global average temperature above pre-industrial levels ought not exceed 2 degrees C. In this regard – we will work between now and (the 15th Conference of the Parties (COP-15) to the UN Framework on Climate Change in December 2009) Copenhagen ... to identify a global goal for substantially reducing emissions by 2050. (Ramachandran 2009)

By contrast, those who believe that India's stand remains uncompromised would insist that the MEF declaration should be read in conjunction with the following statement issued by G-8, which India refrained from signing.

> We recognize the broad scientific view that the increase in global average temperature above pre-industrial levels ought not to exceed two deg. C. *Because this global challenge can only be met by a global response, we reiterate our willingness to share with all countries the goal of achieving at least a 50 per cent reduction of global emissions by 2050, recognizing that [it] implies that global emissions need to peak as soon as possible and decline thereafter.* As part of this, we also support a goal of developed countries reducing emissions of GHGs in aggregate by 80 per cent or more compared to 1990, or more recent years. Consistent with this ambitious long-term objective, *we will undertake robust aggregate and individual mid- term reductions, taking into account that baselines may vary* ... Similarly major economies need to undertake quantifiable actions to collectively reduce emissions significantly below business-as-usual (BAU) by a specified year. (ibid.; emphasis added)

In the light of the above statement the proverbial billion dollar question for many analysts is this: Given that only a global 85 per cent reduction (from 2000 levels) will, as pointed out by the IPCC 4th Assessment Report, have a high chance of preventing a 2–degree increase, will the Annex-1 countries (given the arithmetic based on world per capita emissions) be willing to cut their emissions by nearly 93.3 per cent by 2050? If not (which is more likely), then will China and India (the first and the fourth ranked 'emitters' at present) be prepared to embrace severe limits on their emissions?

The Indian Prime Minister's Special Envoy on Climate Change, Shyam Saran, has emphatically said that 'there was nothing in the (G8) declaration to suggest that India has accepted emission caps'. According to him, 'there can be no contradiction between poverty alleviation, economic and social development and climate change. While flagging India's commitment to ecologically sustainable growth, he argued that India's economy was growing at 8 to 9 per cent annually, whereas the energy consumption was less than 4 per cent. What is particularly interesting is the reassurance given by him that under the NAPCC (released before the G-8 Summit in Tokyo in July 2008), there would also be a massive increase in the forest cover from 22 per cent now to 33 per cent. An additional 6 million hectare of degraded forest would be revived and this would act as a *carbon sink*.

The low per capita emissions argument advanced by India, from which it seems to derive the high moral ground while defending the norm of 'common but differentiated responsibility' in international climate diplomacy, somewhat loses its shine against the backdrop of a multitude of inequities that persist across the country. A Greenpeace-India study found that the carbon footprints of those in the top income bracket in India are 4.5 times that of the lowest (Ananthapadmanabhan et al. 2007). According to Praful Bidwai (2009),

> India's stress on per capita emissions as the sole metric or criteria of equity and the only limit it will accept is problematic. In an extremely unequal and hierarchical society like ours, per capita emissions mean little. They can be a cynical way of hiding behind the poor, whose contribution to emissions is low and hardly rising. It is India's rich and middle classes – which are pampered by the state's elitist policies, and which are consuming as if there were no tomorrow – that account for the bulk of our [India's] emissions increase. There is probably an order-of-magnitude difference in carbon footprints between India's rich and poor.

India has made repeated references to poverty as the key reason for its refusal to take GHG mitigation targets in the ongoing diplomatic negotiations. At the General Assembly in February 2008, the Indian representative said, 'in terms of climate change ... blessed are the poor for they have saved the earth' (Rajamani 2009: 358).

At COP 17, held in Durban, on 10 December 2011, the then Environment Minister of India, Jayanthi Natarajan made a speech, which is said to have received a standing ovation (CSE 2011). Flagging at the outset the fact that 1.2 billion people of India have 'a tiny per capita carbon footprint of 1.7 ton and our per capita GDP is even lower' she expressed her utter dismay over the accusation that India was 'against the roadmap'. Pointing out that India has as many as 600 islands, the Minister reassured the Small Island Developing States (SIDS) that even though India's position was somewhat different, India was sensitive to their vulnerability to the vagaries of climate change. The crux of both the argument and the conclusion was that the principles of equity and CBDR were critically important for India and no dilution of these norms was acceptable to India (ibid.).

At the COP 18 meeting held in Doha in December 2012, India firmly resisted the attempts made by several delegations from the developed world to include agriculture sector in the realm of mitigation. India's argument, reflecting the views of a vast majority of developing countries, was that agriculture should be placed in the context of adaptation, and must not be dragged into the domain of carbon reduction. Sunita Narain (2012), who represented one of India's leading NGOs on environmental issues at the Doha talks, was of the view that as far as the issue of carbon emissions is concerned, the issue of who will cut down and by how much is an issue which needs to be addressed with a sense of urgency. This, however, is easier said than done due to the US stand against the principle of equity and the acceptance of responsibility for historical emissions.

By the time India attended the COP 19 held in Warsaw in 2013, the major finding of the analysis undertaken by the Centre for Science and Education (CSE) was that 'India had lost its space and the momentum in climate talks' and would have to 'go back to the drawing board'. Overall, the talks had little to offer to the developing countries. Whereas some of the developed countries demonstrated a lack of political will to deliver the emission reduction commitments they had pledged earlier, the entire issue of finance – the loss and damage mechanism – remained unclear. At the end of prolonged and tiring negotiations, the Warsaw COP 19 meeting only managed to achieve a 'hollow loss and damage mechanism' without 'clarity on finance by developed countries to assist developing countries in mitigation and adaptation' (CSE 2013).

Critics of India's NAPCC suggest that whereas there are some bold new ideas on paper, such as increasing the contribution of solar energy, the details of how this would be achieved are conspicuous by their absence. As far as the proposed efforts in the direction of a renewed thrust on energy efficiency, an effort to promote integrated water resource management, and a focus on restoring degraded forest land are concerned, once again the devil lies in the detail. According to some critics,

> Much of the plan is simply old wine in new bottles, such as the use of joint forest management committees to 'green India'. Some of these proposals are wine that has long since gone sour, such as the reform of electricity and fertiliser subsidies for farmers. The greater shortcoming is the failure of the NAPCC to articulate a vision, nationally or globally. While espousing a qualitative shift towards ecologically sustainable growth, the plan fails to develop, or even explore, a compelling vision of future development (EPW 2008).

What remains at the core of such contestation are the imaginative geographies of atmospheric space. According to some of the critical perspectives emanating from global South, the already affluent have already filled up the available atmospheric space with pollution and now not much room is left for the rest of the world to grow. Many scientists would point out that the carbon dioxide concentration in the atmosphere has increased from a pre-industrial value of 280 parts per million (PPM) to 379 ppm in 2005. We are further told that the remaining budget is 450 ppm (to keep risks as low as possible) and 550 ppm to be adventurous. The only way the poorer world can take up this remaining carbon budget is if the entire emissions of the industrialized world were to stop now. Will they?

Well within a decade after the signing of the Kyoto Protocol, the emerging political consensus seemed to be that the most effective and efficient way to protect the global climate system is to assign property rights for greenhouse emissions and to trade these rights on international markets. India is a key player in the carbon market today and 'represents a very attractive country' (Nussmaumer 2007) for hosting clean development mechanism (CDM) activities under the UN Convention on Climate Change (UNFCCC). It is significant to note, especially in light of the above statement made by the former Indian Prime Minister, that India's dominance in carbon trading under the CDM is beginning to influence business dynamics in the country. In the month of October 2006, for example, India cornered more than half of the global total in tradable certified emission reduction (CERs).

Under the Kyoto Protocol, and on the basis of the 'common but differentiated responsibility' principle, India is not obliged to cut emissions, as its energy consumption is low. While this may change within a decade or so, companies are jumping on the CER bandwagon. Enterprises are adopting cleaner, sustainable technologies. India Inc. is said to have made Rs 1,500 crore in 2011 just by selling carbon credits to developed-country clients. In the pipeline are projects that would create up to 306 million tradable CERs. According to some estimates, if more companies absorb clean technologies, total CERs with India could reach 500 million (*Indian Express* 2006).

From the standpoint of 'equity' and climate justice it is highly doubtful whether carbon trading will be able to contribute much to the protection of the earth's 'climate' as a global space. To quote Rajni Bakshi (2009),

> Can the sky be owned? And if yes, then by whom? Where should commons end and markets begin? Can value be liberated from the dominance of the price mechanism? How do we decide the value of a 700 year-old tree? We need only to ask how much it would cost to make a new one. Or a new river, or even a new atmosphere. The intrinsic value of the natural world, its right to exist irrespective of usefulness to humans, is fundamentally an ethical matter that cannot be resolved by markets ... Is earth our home or is it one large market place?

According to the Durban Declaration on Carbon Trading, issued on 10 October 2004 and signed by the Indigenous Environment Network (Durban Group for Social Justice 2004), carbon trading is a false solution which entrenches and magnifies social inequalities. The carbon market creates transferable rights to dump carbons in the air, oceans, soil and vegetation far in excess of the capacity of these systems to hold it. Billions of dollars worth of these rights are to be awarded free of charge to the biggest corporate emitters of greenhouse gases in the electric power, iron and steel, cement, pulp and paper, and other sectors in the industrialized nations who have caused the climate crisis and already exploit these systems the most. Costs of future reductions in fossil fuel are likely to fall disproportionately on the public sector, communities, indigenous peoples and individual taxpayers.

The debates on neoliberalism, climate governance and the politics of scale are likely to continue. Ian Bailey (2007) would argue that, 'state acceptance of the principal of collective climate governance, whether by neoliberal or other means, provides few guarantees that commitments will be honoured if these are seen to threaten states' territorially-defined interests.' Whereas according to Jon Barnett (2007: 1372), 'it is the most wealthy people in the most wealthy countries that have the most power to change the political and economic systems that sustain the problem of climate change.' What the state-centric grand narrative of 'global' climate change so ably hides is a 'more subaltern and class-view of climate geopolitics.' According to Barnett, 'The task for a more empowering and critical geopolitics of climate change is therefore to reveal the ways in which climate change is a local and social problem that cannot be solved without the conscious exercise of political and economic choices of people in developed countries' (ibid.). It is in the light of such insights that we turn to the case of China next.

Relentless economic growth, unsustainable environments and persisting climate dilemma: perspectives *on* and *from* China

Despite obvious differences between China and India in terms of their political cultures and political systems, the two fast developing Asian economies, comprising 38 per cent of the world population, continue to face a 'revolution of rising social expectations'. Both are confronted with a multifaceted 'climate dilemma'. Can billions in Asia afford to continue to march on the same path of 'development' as followed by millions in the global North, resulting in rising social-economic inequalities and incalculable environmental costs? Should climate change be approached by China and India as the problem of production or as the problem of consumptions or both? What is more immediate and compelling in terms of priorities and investments: the prediction of a yet to be born threatened mass of undifferentiated humanity or the predicament of the present living generation of millions struggling on the socio-economic peripheries of highly visible, uneven, but situated geographies of neoliberal capital accumulation? As big emitters of greenhouse gases, how will both these countries reconcile the geoeconomic narratives of 'rise', with the claims of millions struggling below the poverty line, and demanding compensation for the 'loss and damage' on account of obvious disappointments with lack of progress over mitigation and adaptation?

No easy answers seem forthcoming to the questions raised above. And yet one might venture that in both China and India – and elsewhere in the global South too – the meaning and scope of complex and compelling phenomena of climate change cannot – and should not – be restricted to 'global warming'. Climate change is better approached and understood in the context of relentless pursuit of economic growth seriously undermining environmental sustainability over a long period of time. Equally significant is the fact that despite break-neck growth in GDP per capita in China during the 1990s – enabling a reduction in the number of the absolute poor from 361 million to 204 million (Sparke 2013: 118) – the in-country disparities between the rich and the poor have increased.

Bryan Tilt (2010) concludes his insightful analysis of the 'struggle for sustainability in rural China' in general and Futian province in particular by pointing out that as PRC enters the fourth decade of liberal economic reforms, 'there are still many important aspects of China's current struggle with economic and environmental sustainability that require further exploration' (ibid.: 162). Given the enormous magnitude of environmental unsustainability China faces – industrial pollution, desertification, deforestation, water contamination, urban expansion, untreated waste, exploding consumption, loss of arable land – Tilt underlines the need for 'scholarly examination of the intricate linkages between ecological and social systems.' In rural China, says Tilt,

> where rapid industrial development has produced pollution problems of immense proportion, a legacy of exposure to environmental toxins and damaged livelihoods remains. Left unchecked, these problems constitute an environmental, social, and public-health experiments whose long-term outcome is unknown but likely undesirable. In Futian, as elsewhere, cadres struggled to find a way to address the government's commitment to environmental sustainability without undoing the economic gains they had fought hard to achieve. (ibid.: 162)

We may use the insights offered by Tilt to map out at some length the state and status of environmental (un)sustainability in contemporary China. Both India and China underline the critical importance of 'common but differentiated responsibility' (and for good reasons in our view) but for various reasons have avoided the internalization of this principle. The moment these 'rising' Asian countries choose to do so, the policymakers would be confronted with the harsh reality of highly differentiated geographies of affluence within their respective societies (even though China can legitimately claim to have pulled up millions more above the poverty line in comparison to India) and the urgency of pushing their climate change discourses beyond mitigation and adaptation. Neither China nor India can put a gloss over the reality that ecological unsustainability of the scale they continue to experience will take a heavy toll over their economic ambitions and performance. Here are just a few facts to illustrate the nature of China's environmental crisis – bordering catastrophe – that, in our view, should be approached and appreciated in conjunction with the challenge of incremental climate change.

Judith Shapiro (2012: 7–10) has graphically outlined some of the facts about China's environmental crisis, with wide ranging implications for human health and well-being. Shapiro cites a World Bank Study to point out that out of the 30 most polluted cities of the world, as many as 20 are in China. The narratives of China's 'peaceful' and 'harmonious' rise – which, as of now, does not appear to be either a peaceful or harmonious relationship with the natural environment – needs to be tempered by the harsh reality that China lost as much as 3 per cent of its GDP in 2004 (out of the 10 per cent growth rate per year between 2000 and 2011) due to economic losses incurred on account of environmental pollution. State Environmental Protection Agency of China (2006) was forthcoming in acknowledging that lack of data did not permit assessment of additional costs resulting from the depletion of resources and degradation of eco-system services. The global media reported the alarming quality of air in Beijing in the lead-up to the 2008 Beijing Olympics with some stories drawing attention to the sorry state of affairs in other parts of China. Two years later, the harsh and painful reality of Chinese 'cancer villages' (nearly 459 cancer villages across 29 of China's 31 provinces) was brought to the global attention by the media and internet (Shapiro 2012:7).

In China, national water pollution is simply alarming; around 21 per cent of its available surface water resources are not fit even for agricultural use. Beyond the impressive labyrinth of rapidly growing cities is the appalling reality that in 2005 nearly half of these urban areas lacked wastewater treatment facilities, leaving untreated, open water and sewage systems (Moyo 2012: 192).

In the case of China, the issue of environmental/climate justice becomes quite convoluted. From one perspective, there are a number of examples to show that lack of inclusive growth in China has marginalized millions. There are millions of slum dwellers, largely rural migrants exposed to various serious health hazards. This class, comprising the 'floating populations', represents the losers of neoliberal globalization in the global South. As pointed out by Matthew Sparke (2013: 117), 'In China rural migrant workers who do not have an urban household registration (known as an urban *hukou*) are effectively reduced to the level of non-citizen in Chinese cities, and this in turn allows for extreme forms of hyper-exploitation and urban inequality.'

Judith Shapiro (2012: 137) would argue that

> China is an importer and victim of environmental injustice, as in Southern Guangzhou, a centre for the international electronic waste recycling business. However, China is also an exporter and source of environmental injustice, as the newly wealthy country displaces environmental harms on to less developed countries, extracting resources from Ecuador to Nigeria and often creating pollution and environmental degradation in the process.

While this is true to some extent, and one finds similar concerns expressed over the Asian and Chinese 'appetites' for food resources against the backdrop of anticipated climate induced uncertainties and scarcities, there is a need for 'critical enquiries into geopolitical representations of food insecurity and of opening media space for a 'counter-geopolitics of food security'', as pointed out by Gong and Billon (2014: 291). Gong and Billon have shown how 'market-driven journalism' continues to frame the notion of 'food security' from the vantage point of national security, invoking the images of 'food riots' in various parts of the global South like Haiti and Mexico. It has of late turned its gaze specifically on 'resource starved' Asia in general and China 'with its rising population and demand for food has been constructed as a food security and environmental threat, generating much neo-Malthusian fear in geopolitical discourses in the West, in particular in the US' (ibid.: 293). Furthermore,

> Securitized narratives of food crisis based on 'neo-Malthusian predictions of an imminent descent into socio-political chaos amidst growing global food supply-demand imbalances' often call forth liberal humanitarian interventions that focus on technical fixes and liberal markets. (ibid.)

China's 2007 National Climate Change Programme (PRC 2007) acknowledged in its foreword that, 'Climate change is a major global issue of common concern to the international community', involving both environment and development, 'but it is ultimately an issue of development.' It was further pointed out that,

> As noted by the *United Nations Framework Convention on Climate Change* ... the largest share of historical and current global emissions of greenhouse gases has originated from developed countries, while per capita emissions in developing countries are still relatively low and the share of global emissions originating from developing countries will grow to meet their social and development needs. (ibid.: 2)

In October 2008, China issued a White Paper on 'China's Policies and Actions for Addressing Climate Change' (PRC 2009) and acknowledged early on that,

> As a developing country with a large population, a relatively low level of economic development, a complex climate and a fragile ecological environment, China is vulnerable to the adverse effects of climate change, which has posed substantial threats to the natural ecological systems as well as the economic and social development of the country. These threats are particularly pressing in the fields of agriculture and animal husbandry, forestry, natural ecological systems and water resources, and in coastal and ecological fragile zones. *Therefore, adaptation is an urgent task for China.* In the phase of rapid economic development, and with multiple pressures of developing the economy, eliminating poverty and mitigating the emissions of greenhouse gases, China is confronted with difficulties in its efforts to address climate change. (ibid.: 1; emphasis added)

Especially since the early 1990s, China's fossil-fuel centric energy system (with coal accounting for nearly 70 per cent) has been going through a transition. As pointed out by Palazuelos and García (2008), one of the outstanding features of this transition is that Beijing has become more dependent on external hydrocarbon markets. In their view, China, the world's second largest economy and manufacturing base for much of the West, stands at an 'energy crossroads'. It is faced with

> electricity blackouts in several cities, obstacles to increasing electricity production in thermal power stations, strong energy demand from industry and urban households, insufficient distribution capacity of the electric power grid, refusal of many mines to provide coal in spite of rising prices, and power plants that lacked fuel while others sold stocks at relatively low prices. (ibid.: 461)

Intervening in the complex energy-environment-climate nexus are a number of the factors including robust growth and changing patterns of industrial specialization, expansion in the transport sector, urbanization and new consumption habits.

Miranda A. Schreurs (2012: 460) has argued that 'Chinese government leaders appear to be serious about their commitment to make environmental improvement albeit on their own terms, not those dictated by Europe, the United States or Japan.' Turning this growing 'commitment' into effective policies continues to be a rather daunting task in a top-down, centralized party-State like China. Conversations with the local communities on the one hand, and governmental officials on the other, reveal a good deal of uncertainty, bordering on confusion, over the nature, scope and reach of 'governance' and policies. Moreover, as pointed out by Judith Shapiro (2012: 59),

> Competing and overlapping bureaucracies plague every level of administration, and the lines of authority from the 'centre' in Beijing to the localities are often weak, with government officials at the prefecture, district, country and township levels often answering to other local officials rather than to their superiors in the central environmental bureaucracy.

Consequently one finds divergent, competing views on ecologically sustainable development. 'Within the bureaucratic system there are strong advocates for a new, more ecological form of modernization, but they face serious challenges' (ibid.).

A number of studies do suggest however that much needed space for civil societies to function and popular collective action to protest has somewhat expanded in recent past. A recent study by Chen (2012) aims at answering the following puzzle: 'why there has been a dramatic rise in, and routinization of, social protests in China since the early 1990s?' Chen draws attention to as many as 9,213 collective petitioning events (i.e. petitions delivered by five or more participants) in Hunan province in 2001, taking place in the county-level government or higher, with some resorting to 'troublemaking' tactics such as highway blockades, demonstrations and sit downs. His argument runs as follows: in the case of China it is useful to bear in mind that, far from being a 'monolithic authoritarian regime' or a 'unitary state', the PRC is a state (rather Party-state) that is beset with divisions and contradictions. What facilitates mass mobilization and affects popular contention therefore are not only the 'vertical divisions within the Party-state – that is divisions between the central government and local government' – but also the 'horizontal divisions, such as those among different Party-state agencies at the same level' (ibid.: 15). To quote Chen,

> In the Reform Era, many public agencies have begun to seek a somewhat independent identity and are acquiring distinct institutional interests. To obtain either institutional power (such as in the case of the Peoples' congresses and mass organizations) or commercial interests (such as in the case of the media), these agencies often take a position somewhat inconsistent with that of the Party and government leaders. *Coordination among different agencies has thus become increasingly problematic*. This differentiation has to some extent created the multiplicity of quasi-independent centres of power within the regime. Therefore such divisions and differentiations not only help create grievances, but more importantly, they also produce multiple allies and advocates for petitioners and therefore offer invaluable resources and protection to them. (ibid.: 15–16)

A critically important point (with significant implications for the rhetoric and reality of 'sustainable development' and 'climate change') made by Chen is that one needs to acknowledge the 'contradictions and ambiguities between state institutions and ideology' (ibid.) and appreciate that these provide yet another source of political opportunity to voice and wage a social protest.

Conclusion

Climate change, as a site, as a discourse, as a form of territory – call it what you will – is a product of the North, most particularly Western Europe. It is a story embedded within post-industrialism; it is a story well-suited to theoretical developments in post-modern theory and post-political theory; it is a story well-suited to a moral conservatism, whilst simultaneously advocating and lionizing a neo-liberal global subject. In this globalised nothingness, this imagined transnational *non-territory*, it is dreamt that climate change is an opportunity for co-operation. As Chesters and Welsh (2006) write, boundaries fall away, as the potential coalitions of interests multiply as the once firm boundaries constituting social groups and actors are subject to increasing rapid perturbation as 'All that is solid melts into air' (ibid.: 5).

But, of course, this makes little sense in the context of the suffering – the realm of the global South. There are many parts of the planet where corporate-led globalisation has not won yet. As one of us has written elsewhere:

> It is useful to draw a distinction between the 'objects' and the 'subjects' of the emerging world order. Geo-politically speaking, the

> world order is not as 'global' in nature, scope and functioning as many of us believe it to be or would like it to be. In other words, emerging order is not as *placeless* as those who are fascinated by the time-place compression of globalization would like to imagine ... place still matters for production, reproduction and consumption. (Chaturvedi and Painter 2007: 388)

Of course, the separation between North and South is a useful category marking out the affluent lives of the minority from those of the less-affluent majority. But, like any border, ultimately it is drawn subjectively and imperfectly to demarcate territories (in this instance by us). As illustrated by the Indian case, the line between Northern support and Southern rejection of the climate agenda is not, in real terms, so absolute. Although in the first years of climate's appearance on the global green agenda, many Southern environmentalists rejected it as a form of green imperialism. But now, as we will discuss in Chapter 7 with reference to labor, religious and green movements, climate's staggering breadth of ideological reach has re-mutated into versions of the climate discourse, which include environmental justice arguments. In this light, climate justice does attempt to grabble with notions of environmental debt caused by centuries of ongoing colonialism.

In short, by now re-territorializing all major environmental issues into one climate category, climate security is a flawed position on two counts: first, environmental catastrophe for the many in the global South is a daily reality, not a calamity-in-waiting. Secondly, the ultimate day-of-judgment, a future day when the earth's climate change will lead to another great flood, imagines an environmental punishment being dished out, ultimately by forces of nature. Projecting a force-of-nature as the ultimate source of retribution conveniently provides cover for the key perpetrators, mouthing climate change platitudes from their homes and universities in the affluent world. What is glossed over by these imaginative geographies of climate change, 'global' as well as 'national', is the long-standing history of a multitude of socio-ecological injustices in the global South. As Amartya Sen has persuasively argued in his recent book entitled *The Idea of Justice* (2009: 26),

> Indeed, the theory of justice, as formulated under the currently dominant transcendental institutionalism, reduces many of the most relevant issues of justice into empty–even if acknowledged to be 'well-meaning' rhetoric. When people across the world agitate to get *more* global justice ... they are not clamouring for some kind of 'minimal humanitarianism'. Nor are they agitating for a 'perfectly just' world

> society, but merely for the elimination of some outrageously unjust arrangements to enhance global justice ... and on which agreements can be generated through public discussion, despite a continuing divergence of views on other matters.

The insights offered by Amartya Sen are equally relevant with regard to the notion of 'climate (in)justice'. Furthermore, advocates of dominant, Northern climate change discourses (in their selling of a global soul to the global South) dismiss concepts of *intra-generational equity* (the rights of the legitimate other) and seek to replace them with the post-material concept of *inter-generational equity*.

In this chapter, we have depicted climate as an issue which deterritorializes existing geopolitical realities in a manner which suits the discourses of both elite science and corporate globalization. In this deterritorialization, the politics of place, of difference, are removed; the divisions between North and South – the Minority and Majority Worlds – must melt away as all peoples become citizen-consumers in need of a morally conservative (using global archetypal myths of flood and fire) but economically neo-liberal *Global Soul* with which to confront the global nemesis of climate change. This deterritorialization is constructed from a Northern (particularly a western European) position. It emerges from post-material and post-industrial environmental discourses, largely ignoring the discourses and frames of post-colonial environmentalism (and environmental debt) which are far more appropriate when describing the environmental and developmental realities of the global South.

4

The Violence of Climate 'Markets': Insuring 'Our Way of Living'

Introduction

Larry Lohman (2008: 364) has persuasively questioned the belief that climate justice is all about 're-energizing or reforming development and investment in the global South to steer it in a low-carbon direction, harnessing the potential of carefully constructed green markets, or making capital flow from North to South, instead of from South to North'. To do so, argues Lohman, amounts to putting a gloss over the 'lessons gained from more than a half-century's popular and institutional experience of what development – neo-liberal or otherwise – actually does.' Lohmann rightly asks: 'what does the project of a just solution to the climate crisis become once it is associated with or incorporated into an economic development or carbon market framework?' To quote Lohmann, '... carbon trading as part of the "climate development" package that has become entrenched at national and international levels over the past ten years, is organized in ways that make it more difficult even to see what the central issues of climate justice are, much less to take action on them' (ibid).

Of course, there are other responses to climate change which are not neo-liberal. But, no doubt, it is the neo-liberal market responses which are most dominant. Carbon-trading, as aforesaid, is the most obvious form of the neo-liberal response, as it promises to deliver us to a carbon-neutral future by trading between the affluent world and the less affluent world. In this vein, the global South sells its future capacities to produce carbon (and to industrialize) to the North, which not only continues to produce emissions, but, in real terms, increases them. The North then uses these carbon offsets not for business-as-usual; but for business *better-than-usual*.

The key attribute of the climate story remains in its ability to ride over, and move through cultural and religious boundaries through its use of key archetypal mythologies built upon the solid rock of shared humanity, the base points of flood and fire which inform and cross over the boundaries of history. By eradicating difference at this most primordial level, using morally conservative stories which challenge little, a global subject is now made available to serve as a malleable global consumer, one that is perfectly placed to play her part in the worldly game of neo-liberal capitalism.

Climate crossings on the right-left political spectrum: but born of the right

Due to its wide cultural reach and conservative core (it challenges little regarding 'the order of things') the climate story is also unusually well-placed to serve a neo-liberal economic agenda. Environmental issues which emerged within the first flush of modern environmentalism demanded concepts such as limits-to-growth. Environmental issues were portrayed as a zero-sum game, with trade-offs demanded of business in order to achieve environmental goals. During the mid- to late-1980s, sustainable development discourses (neatly attached to the ideology of ecological modernization) began to replace the finite growth hypothesis, and we discovered that business was, in fact, good for the environment and, of course, environment was good for business. In terms of sustainable development and ecological modernization, nation-states were still seen as responsible to provide, in classical liberal fashion, the role of responsible regulator, legislator and monitor of the anti-social practices of a minority, to protect the interests and environments of the majority. Since the mid-1990s, a new depiction of the politics of environmental concern has become dominant; one that shares much with its sustainable development cousin, but is far more extreme and brutal in its embrace of market-principles. These recent manifestations are sometimes known as 'wise use', 'sequential use', or most obviously, 'neo-liberal environmentalism'. Whereas sustainable development turned the win-loss game into a win-win game (business-as-usual), neo-liberal environmentalism takes the next step, constructing a win-win-win game: business better-than-usual. Climate change, and the cacophony of issues which gather at this site, is the purest form of this win-win-win construction of the environment.

The omniscience, the power of the climate change debate has meant that nearly all positions occupying the ideological continuum in traditional political economy have reacted to its specter. A key attribute of a political myth is its aforementioned ability to rally a whole range of diverse (and sometimes disparate) beliefs within its single domain (Edelman 1978). It is used for a cacophony of political purposes. Climate change does this beautifully, crossing the complete political spectrum, and providing positions which suit those on parts of Left, to others fashioned for those on the most rigorous branches of the Right. The Left positions, for example, of climate justice economics have, however, emerged later largely due to 'issue capture' and for the purposes of political opportunism, not for any inherent qualities of the climate change debate as they relate to delivering leftist policy outcomes.

As will be argued later in this chapter, one of the central reasons why the neo-liberal right has captured the climate debate so easily and so successfully is that the climate change scenario has always been based on a right-wing (particularly conservative and 'realist') ideological and moral position. As discussed in Chapter 2, the 'Armageddon', 'catastrophe' and 'scarcity' climate science positions are very much rooted in political conservatism. And, like society, environmental movements possess activists from all parts of the left-right political and economic spectrums. Anthropogenic climate change, in its early manifestations, was most often championed by greens on the right.

In short, its central tenets are borne of right-wing ideology and mythology, though the more extreme, more recent neo-liberal manifestations are new. So, it has not taken much for neo-liberal, industry-funded, *wise use* think tanks like the Lavoisier Group, Frontiers for Freedom, Clean Air Institute, Environmentalists for Nuclear Energy, and the Clean and Safe Energy Coalition, to re-position their arguments in a manner which takes them from their previously popular realist and economic nationalist positions of *climate denial*, to their new position of *climate co-option*. Born of the right, climate change – regardless of the warm, fuzzy re-interpretations of the Left (who have tried to co-opt, largely unsuccessfully, the climate position for their own political purposes to include, as we shall see, concepts of *climate economic justice, climate debt, climate redistributive economics* and so on) – will live on and die on the right, and be dominated by these more neoteric right-wing depictions of climate change and its market-based 'solutions'.

In the rhetoric of international relations and international political economy, climate change was always positioned equally well within the frames of mercantilism/economic nationalism, liberalism, neo-conservatism and neo-liberalism – all ideological positions derived from the Right. It first emerged as a powerful item on the agendas of the affluent, minority world in the 1980s when it was finally understood that regardless of nation-state boundaries, the wealthy could not insulate themselves from the rest of the world; that they had to share the earth's survival systems with the multitudinous poor; and measures were then sought to control the poor, as the poor were reconstructed as the most environmentally degrading people/threat on the planet (Doyle 1998). In this vein, climate security (see Chapter 6) became part of *global political ecology/economy*, which, in turn, became part of *global security*.

The cross-overs between geo-economics and geo-securities are deeply embedded in our international systems, and for this reason their interconnections are further explored in Chapter 6. But for now, all that needs to be stated on this subject is that in reality, global climate and ecological geo-economics and geo-securities are really *national* securities for societies and markets of the more affluent, minority world nations. In an article which assesses the national security implications of environmental threats in the United States, including climate change, Marc A. Levy writes:

> For action on problems like climate change, however, we need a policy-making style more like defense policy than environmental policy. Climate change is a problem much more like the problem of containing the Soviet Union; it requires a grand strategy to guide actions in the face of distant, uncertain threats, and an overarching commitment from high levels of leadership to stay the course through the ebbs and flow of popular sentiment. (Levy 1995: 54)

Why was the climate change position born of the Right? Surely, defenders of climate change will argue that the position came to the fore out of necessity; that climate change is *real*; that the ecological crisis has led to the political reality. In part, this is correct. Of course, one of the earliest and most popular positions on the Right has been that climate change does *not* exist; that it was simply a scientific position dreamed up and then crafted by environmentalists (particularly the right-wing variety of Green) to challenge business-as-usual, to restrict industry and development interests from making profits which were rightfully theirs. This is the 'climate denial' position. Industry-funded, right-wing think tanks – such as the Heartlands Institute, and the Institute for Free Enterprise – were particularly active in both Australia's and the United States' eventual decision to withdraw support from the Kyoto Protocol process. In research aimed at monitoring and analyzing these think tanks' impact on the United States' climate change policy, McCright and Dunlap noted that: 'the conservative movement employs counter-claims to block any proposed action on global warming that challenges its interests' (McCright and Dunlap 2000: 518). Despite some of these claims being clearly nonsense – such as, the Kyoto Treaty 'would cut economic growth by 50 per cent by the year 2005' – they were backed by large public relations concerns with almost unlimited financial backing from multinational extractive industries. According to McCright and Dunlap, despite these arguments' lack of veracity, they 'played a decisive role in defeating the U.S. ratification of the Kyoto protocol' (McCright and Dunlap 2003: 367), proving that powerfully backed political myths can beat scientifically 'correct' ones.

Because of the past dominance of this anti-environmental position, in the early years of the climate issue's emergence, it made it very difficult to air views which critiqued the climate position within more centrist and leftist environmentalist circles (whatever the basis of this critique) as they were inevitably cast as supportive of 'the enemy', and therefore 'radio silence' was maintained at all costs. Of course, this is not new to the climate debate. Most issues are portrayed in the popular press as dualistic – either for or against – whether it be evolution versus intelligent design; abortion versus right-to-life; anti-nuclear, or pro-nuclear etc. Also, environmental movements often strike out mistakenly at diverse positions within their own movements. Rather than understanding the mechanics and organics of their diverse and amorphous political form, they mistakenly and increasingly discipline themselves along the lines of more unidimensional political parties and corporations, demanding 'one voice' from their members (Doyle 2001).

Make no mistake, as the 21st century moves towards its third decade, the climate denial position still has numerous powerful allies, but many of those who still advocate this line now choose to advocate it in more private circles. Now, their public utterances have become more sophisticated, aided and abetted by right-wing think tanks (which themselves have changed track) helping them to 'green their products' for the purposes of 'improved positioning in the marketplace'. This can be called the 'climate co-option' position: 'we don't believe it ... but, if we can't beat them, join them, and then beat them at their (the environmentalists') own game'.

In Chapter 7, we move to the more recent green economic and political climate responses of the Left which, although nowhere near as prominent, deserve mention at least in the manner in which they have responded to the Right's capture of the climate change economic debate. We argue that in the 21st century, the Left – in relation to debates pertaining to the political economy of climate change – is hardly breathing. All that remains are occasional debates between the factions of the Right themselves, with the neo-liberal 'carbon market' responses being by far the most vociferous and the most effective in terms of agenda dominance and policy delivery. Climate change, as an issue, has further entrenched the truth that the Earth is not a society, but purely an economy.

For the body of this chapter, therefore, we first list the dominant ways in which the Right has now positively interpreted climate change using the language of political economy as a tool, and then providing working examples of these approaches. In the last part of the chapter, we look closely at by far the most dominant right-wing political economy position: neo-liberalism and climate change.

The economic nationalist (Mercantilist) response to climate change

On the right, within the mercantilist or 'realist' understandings of political economy, the role and importance of the state is obviously accentuated. Key assumptions of this position are based on: the primacy of the group (in this case the modern state) rather than the individual over most elements of social life (O'Brien and Williams 2010). Also, critical to this position's premise is Hedly Bull's contention that the inter-state system is anarchical and it is therefore the duty of each state to protect its own interests (Bull 1977). In this vein, the state is the pre-eminent actor in the domestic and international spheres, existing *a priori* to the market, and the international system is a struggle for power and wealth between rational states. Relations between states are characterized by unending conflict and the pursuit of power, and the *nature* of the global economy reflects the interests of the most powerful states. In this mercantilist view, due to the fact that markets can oftentimes be negative in their effects, state control of key economic activities or, at least, state assistance to central economic sectors is favored. Security concerns dictate that too much dependence on key energy sources derived from abroad is undesirable, despite any economic benefit (O'Brien and Williams 2010). In this manner, from an economic nationalist view, the United Kingdom, for example, may embrace alternative energy systems, not only because of climate change being viewed as an *essentialist threat*, but also, for example, to guard against an over-reliance on Russian gas and oil, which may threaten its national security in the future.

On most occasions, the economic nationalist position has often been used to quash climate change imperatives, as they are usually constructed as a 'beyond the state' or 'transnational' problem. For example, the United States' refusal to sign the Kyoto Protocol was almost totally based on the fact that its government regarded the agreement as not in 'the interests of its citizens'. But, in other instances, let us not forget that these forms of neo-mercantilism can also be used not to deny the climate change imperative, but to champion the rhetoric of climate change to equally protect a particular nation's place in global finance and trading systems. So, on the right, climate change has also appealed to the economic conservatives. These nation-state based mercantilists can articulate positions of comparative and 'natural' advantage either to reject climate change, or to embrace it. For example, why would countries such as Australia, South Africa or the United States, rich in their fossil fuel deposits, jeopardize their market advantage by moving away from these traditional sources of energy? On the same side of the coin, why wouldn't fossil-fuel poor countries such as the United Kingdom, France or India articulate the need, for example, for nuclear technologies (often portrayed as more 'carbon-friendly' than coal or oil) or other non-carbon based fuels in order to satisfy their economic nationalist agendas?

These economic nationalist positions are also built on political conservatism. Bull's anarchic international system is also mirrored in the natural order itself. Nature is at war with itself, in a state of 'natural chaos', and order – whether in the international system or in nature itself – can only be brought about by nation-states acting in their own interests. One of the most obvious instances of the conservative green economic position response relates to 'scarcity economics'. This Malthusian line of reasoning has been used to justify birth control programs and to urge no response to poor people in need as a result of famine and drought. It was an extremely popular argument in the 1960s and 1970s, championed by both Garrett Hardin and Paul Ehrlich. Their argument begins with the assumption that there are already too many people consuming the Earth's limited resources. A 'population bomb' is set to explode. With such a narrative other questions emerge: what is the Earth's 'carrying capacity'; and who should have access to increasingly scarce resources? (Doyle and McEachern 1998: 71.) The debate about 'carrying capacity' within green movements is an essentially conservative one. It assumes 'natural limits' placed on earthly existence by some form of transcendental moral authority: 'Nature' with a capital 'N'. The conservative green economic response, in this case, has largely been about reducing absolute numbers of human beings, usually from the less affluent world. Poverty *per se* is seen as one of the main causes of environmental degradation, not a result of it. There is still much talk in the 21st century of domestic 'carrying capacity' using climate change arguments to justify a reduction in total human numbers, particularly within the global North (meaning no more immigrants).

So, again, the climate change debate has to be understood as a sub-set of these environmental positions. In this specific carbon-centric case, it is obvious that the scarcity argument is aimed directly at fossil fuels, and that ultimately this lack of resources to match population growth will lead to problems with *environmental security*, as 'the blood-dimmed' tides of the majority world will sweep away the lives of the wealthy minority. This scarcity argument is often directed at countries with high populations, such as India and China. Very simply, the economic argument is that these large industrializing populations will destroy the planet if they continue to develop using carbon-based technologies.

In Chapter 2, we listed climate change's conservative appeal to discourses of catastrophe, chaos, apocalypse and judgment. Both Crist (2007) and Hulme (2008b) note the apocalyptic nature of climate change discourse. In the case of the former, she observes that this discourse and language plays into the hands of the religious fundamentalists of the world, and is closely aligned with the old Testament narratives of the end of the world (Crist 2007: 47). Hulme also refers to these connections with scriptural readings. There are limited naturalistic understandings of weather, and in their absence, theological understandings of the relationship between God and nature become dominant. There are issues of divine retribution for moral failings, the workings of Satanic forces such as witches and demons. Hulme provides three examples from the 16th to 18th centuries to illustrate this point (Hulme 2008b: 7–8). This dominant framing, he argues, becomes more and more diluted as of the 18th century as naturalistic explanations for climactic phenomena become available. Note, however,

> Yet traces of this narrative remained, as in later vigorous Victorian disputes about the relevance and efficacy of prayer for stopping or starting rain (Turner 1974). And hollowed-out theological orientations towards explanations of extreme weather can still be found today, whether in the linguistic convenience of 'Acts of God' for the insurance industry or in the theological repertoire ('sin', 'guilt', 'penance') of contemporary discourse around individual carbon footprints and climate change. (ibid.)

Climate change as *judgment* is of course closely related to climate change as *catastrophe*. Again, Hulme neatly traces the history of this discourse which, in its earliest manifestations (early 20th century), considered climate change to be benign or beneficial (2008: 10). Anxiety grows and the discourse then switches from fears of nuclear winter to global warming. In this vein, he sees fears about climate change taking the place of fears of nuclear destruction during the Cold War, and then receiving an extra impetus as a result of the 'war on terror'. Hulme regards the conservative language being used now as having a 'quasi-religious register' (2008b: 11), given its apocalyptic qualities. The sense of emergency is absolute, and the only economic response is a neo-mercantilist one, in order to protect the interests of the nation-state operating in this anarchic and catastrophic climate system.

In more recent times, these catastrophe tropes have been written large in discourses of *tipping points*. Faced with the enormously difficult task of marketing incremental, abstract climate change to its audiences in a photogenic frame, the environmental organization Greenpeace decided to name its 1994 report 'Climate Time Bomb', and used the visual imagery of a nuclear holocaust – a setting or rising sun being suitably edited to look like the mushroom cloud of a nuclear warhead explosion – to communicate the sense of urgency.

Some of the earliest deployments of alarmist metaphors were by the scientists themselves. For example, James Hansen in his writings and pronouncements during 2004 used expressions such as 'time bomb', 'Humanity's Faustian Climate Bargain', and spoke of a 'slippery slope to climate hell', while being highly critical of objections that were raised to such uses. Hansen justified his experiments with new metaphors on the grounds that dangers associated with climate change had to be conveyed to both the governments and the public at large as a matter of highest priority. Hansen (2005a) explained his reasoning in the following words: '"A slippery slope to Hell" did not seem like an exaggeration. On the other hand, I was using "slippery slope" mainly as a metaphor for the danger posed by global warming. So I changed "Hell" to "disaster"' (ibid. 269). A year later Hansen first introduced the metaphor of a 'tipping point' in his 2005 address to the AGU, and received extensive media coverage. What Hansen implied by the usage of this terms that, 'we are on the precipice of climate system tipping points beyond which there is no redemption' (Hansen 2005b: 8).

In his book strategically titled, *Storms of My Grandchildren: The Truth About the Coming Climate Catastrophe and Our Last Chance to Save Humanity*, Hansen placed sea level rise as the *most* dangerous climate impact 'because the effects are so large and because it would be irreversible on any time scale that humanity can imagine' (Hansen 2010: 144). The other climate change impact at the top of Hansen's 'dangerous list' is 'extermination of species'. According to Hansen, 'Survival of both ice sheets and species both present "nonlinear" problems – there is a danger that a tipping point can be passed, after which the dynamics of the system take over, with rapid changes, which are out of humanity's control' (ibid.). The rhetorical links between climate and nuclear catastrophe are many.

The nuclear response: climate catastrophe depicted worse than a nuclear catastrophe

One of the interesting changes in the climate debate on the Right has been the re-emergence of nuclear energy as *solution* to the problems of carbon-based energy. As will be touched upon in Chapter 6, this is in large part a result of the securitization and militarization of climate change issues. But, the re-emergence of nuclear energy is also an economic response to the issues at hand. There are two main reasons why the selling of nuclear energy is so attractive to mercantilists. First, the catastrophe scenarios deployed by conservatives listed above can continue to be deployed in ideological terms – in traditional ways of maintaining authority and control of the populace. Secondly, and most importantly, nuclear energy is largely controlled by the state. This is extremely attractive to economic nationalists. This is not to say large corporations are not involved in the production of nuclear energy, but rather, it is usually done in close conjunction with the state, and ultimately controlled (and often owned) by the state. The reasons for this are rather obvious. Nuclear energy cannot be decentralized in the manner that solar, wind-generated energy, or even carbon-based energy can be. Nuclear energy is both power in terms of energy, and power in terms of state-control.

One spin-off from this process has been the too-clever-by-half repositioning of the uranium mining and nuclear industry as one which not only accepts the possible ravages of global climate change, but seeks to provide a 'renewable', 'alternative' form of energy which is 'greenhouse friendly'. This little public relations side-step and jig has been greeted with rapturous applause from the conservative right, supported by huge mining interests, state and federal governments alike (of both major party political persuasions), as they seek quick moral justification for selling uranium to the highest bidder, regardless of those bidders' standing under international law; whether or not they have signed the United Nations weapons non-proliferation treaty; whether or not they have adequate safeguards in place in relation to potential reactor meltdown or other malfunctions in the nuclear fuel cycle; whether or not they have adequate provisions for the treatment of nuclear waste; whether or not they are likely to use uranium to construct WTMDs: Weapons of Truly Mass Destruction.

Now this 'green' position is also vociferously supporting the 'cradle-to-grave' options for the nuclear industry – digging up the uranium; converting it into yellow cake; selling it to end-users; and then taking the waste back for storage (again for the highest of moral principles). No doubt the gaps in the nuclear cycle will soon be advocated over the coming years, including long-time nuclear energy-shy Australia, which is positioning itself to becoming a nuclear energy producer, with the construction of more nuclear reactors on home soil. Even the Australian Labour Party (once the hand-brake to the nuclear industry in its own country) abandoned its 'three mines policy' in time for the 2007 Party Conference, ably assisted by income-hungry state Labor Governments.

In the United States, conservative think tanks within the Wise Use Movement first saw the opportunity to revive pro-nuclear sentiment in the early years of the new millennium. Patrick Moore, a wise-user who has worked for the logging industry in the US, conveniently referred to as Founder of Greenpeace, actively pursues this style of rhetoric. In his website entitled 'Greenspirit' Moore co-opts green and democratic language to further extractive industry's gains. In recent times, Moore has now taken up employment by the nuclear industry, setting up the Clean and Safe Energy Coalition (Wald 2006). This particular 'climate co-option' argument – that nuclear energy is green, clean energy – is now commonplace. The Nuclear Energy Institute (the main nuclear industry organization in the US) hired Moore to engage in 'grassroots advocacy'. In a statement to the US Congressional Committee in 2005, Moore made the climate/nuclear link the primary argument of his presentation, one which continues to inform his wise-use style, public relations work for the industry: 'A significant reduction in greenhouse gas emissions seems unlikely given our continued heavy reliance on fossil fuel consumption. An investment in nuclear energy would go a long way to reducing this reliance' (Moore 2005).

Again, in Australia, Moore has been actively greenwashing big business interests. He was selected by 60 Minutes reporter Peter Overton to provide the 'balance' to his story on 'The Nuclear Solution'. Moore was identified for the purposes of the story by using his ex-Greenpeace credentials, but his affiliation to the powerful nuclear industry-funded lobby group was not (Overton 30 April 2006). Again, the climate change argument was the central premise to Overton's story, with nuclear energy posed as the key potential solution.

By 2010, these pro-Nuclear public relations campaigns were now gaining significant traction in US government circles. Russell and Hunter (2010) discuss the enthusiastic financial support provided by the Obama administration to the nuclear industry using carbon-friendly credentials. They report that a $US8.5 billion U.S. loan guarantee was provided by the Department of Energy in February 2010 for two new units at the Vogtle nuclear plant in Burke, Georgia, as part of the nation's clean-energy future. In addition, the President's 2011 budget included $US36 billion for half-a-dozen new nuclear plants (ibid.). In more recent times, despite the clean, green image of the nuclear industry remaining in the US, cheap natural gas extracted within the United States from the highly controversial process known as fracking has taken the edge off the domestic nuclear revival in North America (Smith 2012).

In countries where continued cheap domestic carbon-energy extraction is not an option, the nuclear 'green' climate option has sustained its momentum. In a fascinating article documenting the connections between climate change and nuclear energy debates in three European countries – Finland, France and the United Kingdom – the pro-nuclear position is being further entrenched but for a range of cultural reasons (Teräväinen et al. 2011). In Finland, people are largely willing to go along with the 'clean nuclear' option on the basis that 'technology-and-industry-know-best'; in France, it's a case of 'government-knows-best'; and in the UK, most people are willing to concede the use of nuclear energy as a carbon-neutral alternative based on the premise that 'markets-know-best'. In all three countries, heavy investments in nuclear energy are being continually justified on the basis of avoiding carbon-based catastrophe, and setting up alternative, non-fossil fuel nationally-controlled energy sources and grids. This is economic nationalism at its most rampant using climate change as its cause celebre.

But the most remarkable 'capture' of the climate change position by pro-nuclear sentiments has occurred within the green movement itself. The Environmentalists for Nuclear Energy webpage promotes green movement activists who have now seen the nuclear powered-light as a means of averting climate catastrophe (EFN 2012):

> Since EFN was created, many of the greatest environmental leaders have joined our organization or share our views such as, for example,

> Pr. James Lovelock (author of the Gaia theory), Patrick Moore in North America (co-founder of Greenpeace international in 1971 in Vancouver, Honorary Chairman of EFN-Canada). Others include: Stewart Brand (author of the Whole Earth Catalogue), late Bishop Hugh Montefiore (founder and former director of Friends Of the Earth until he joined EFN) and Stephen Tindale (former executive director of Greenpeace U.K. from 2000–2005) who protested against nuclear power for 20 years, but changed his view in 2009 to support clean nuclear energy & has founded the group Climate Answers. The Environmental Defense Fund (EDF) which was hostile to nuclear power until 2005, is now neutral to it. The environmental group Center for Environment, Commerce & Energy (CECE) supports nuclear power in the US.

This pro-nuclear turn has been particularly pronounced in the United Kingdom (The Energy Collective 2012). Chris Goodall (2009), a Green Party activist and parliamentary candidate, in an article entitled 'The green movement must learn to love nuclear power' (Goodall, *The Independent* 23 February 2009), writes from this new-found green perspective, using carbon-based moral imperatives for justification for the switch:

> This country faces a serious energy crisis. Within a decade a large fraction of the UK's antiquated power-generating capacity, both coal-fired and nuclear, is due to close. If it is not replaced, we face a nightmarish future of power shortages and blackouts. In the meantime, we desperately need to reduce this country's greenhouse gas emissions: 90 per cent of our energy currently comes from fossil fuels. This country's current and past emissions are far more than our share of the world population. Unless we reduce our carbon pollution urgently, we will be in breach of our moral, as well as EU and UN, obligations ... This is particularly the case for nuclear power (ibid.).

This focus on national economic energy goals, whether they be broadly green, carbon-friendly or otherwise, has fallen directly into the hands of conservative national governments, eager to continue to pursue nuclear energy and nuclear security for economic nationalist ends.

The liberal response – Kyoto and beyond – towards a climate-friendly, green welfare state?

Also on the right is the economics of liberalism. In this category, the state is still important but is not regarded as an homogenous actor, with a whole range of other contributing political partners emerging to the fore at times, from corporations to interest groups. Liberalism as a theory of political economy emerged forcefully in 18th and 19th century Britain in the wake of the industrial revolution. There can be no doubt that ideological variants of liberalism still dominate the running of the global political economy in the first part of the 21st century (O'Brien and Williams 2010). On a sliding ideological scale, most classical liberal economists (or liberal institutionalists) still champion the primacy of the state (with its networks of interdependence both within and outside the state); whereas neo-liberal, or free-market economists regard the state as diminishing in an emerging post-political, largely borderless global economy, dominated by large, Multi-National Corporations (MNCs) (Gamble 2009). In our work, we regard this latter strand of neo-liberalism as becoming so dominant in discussions of climate change economics, that it has taken on the mantle of an ideology in its own right. So although we acknowledge the classical liberal traditions of neo-liberalism, for the purposes of discussion, we will treat them as separate categories.

The liberal tradition views natural processes, including human nature, in a substantially more generous light than their conservative cousins, and this has a profound impact on the kinds of economic responses liberal thinkers utilize in the market-place. Classical liberal theorists like John Locke and Jeremy Bentham argued that the vast majority of people are 'good'; but they are sometimes hampered by an anti-social minority. So nature, in this light, is not in a perpetual state of war, but can get into minor skirmishes from time to time. In a bid to protect the majority from the anti-social minority, liberal theorists argue for the establishment of a moderate state, not an authoritarian Leviathan, to promote the interests of the majority. Although liberalism does have some notion of 'greater good' it is still heavily based on the rights of the individual, attempting to allocate autonomy to each citizen within responsible bounds umpired by governments. Liberalism is heavily reliant on private property and is closely associated to the politics of capitalism with its emphasis on continuing economic growth. It is also based on pluralist assumptions that all citizens have equal access to power and resources, often denying inequities based on class, gender, race and species.

The liberal green economic response to climate change is sometimes designated as 'shallow ecology'. Most obviously, it has led to the establishment of laws and legislation which provide environmental protection, while not interfering unduly with the everyday mechanics of capitalism. One instructive example of a liberal green economic response was the development of much environmental legislation in the 1970s in the global North, and also the establishment of Environmental Protection Authorities (EPAs). The EPAs and other such authorities were established to monitor the excesses of an environmentally degrading minority and to punish those through a series of fines, impositions and further regulations. In market terms, individuals pursuing 'enlightened self-interest' will maximize the benefits of economic exchange for society. In this light, international relations are seen as 'essentially pluralist, interdependent, cooperative and peaceful, at least to the degree that capitalism and free markets dominate' (O'Brien and Williams 2010). But what is critical about liberal approaches to climate change (and what differentiates them from the neo-liberal) is the central place of the state. The question is then asked: Can the state – or an international community of states – deliver a form of governance which can adequately address climate change now and in the future? Can global summits like Kyoto and Copenhagen provide sufficient co-operative, state-based governance to save the Earth from climate crises, in terms of providing agreed upon measures for both mitigating and adapting to climate change?

Since the modern environmental movement emerged in the late 1960s, increasing scientific information and more active lobbying by environmental parties and pressure groups have combined to produce a growing ecological awareness in the citizenry of many states. As a result, the past two decades have witnessed the development of more coherent strategies in states, in part to meet commitments undertaken as part of international agreements such as the Montreal Protocol (1989), Rio (1992) or, in specific climate change agreements such as Kyoto (1997), or as the result of concerted lobbying action by green social movements (Dryzek et al. 2003). Smaller states in Europe such as the Netherlands, Sweden and Norway have moved more rapidly to integrate environmental concerns with core state functions (Meadowcroft 2005: 4). Indeed, most developed states have seen the development of specific ministries or agencies dedicated to the environment, as well as national action plans to guide future efforts (Catney and Doyle 2011a).

While many environmentalists considered states, at best, as a blockage to environmental protection and, at worse, to be playing a critical role in generating environmental crises, more recent contributions have sought to 'bring the state back in'. For example, Eckersley (2004), Meadowcroft (2005), Doyle and Doherty (2006) and Giddens (2009: 5), all highlight the essential role that the state will play – and in some cases is already playing – in structuring responses to climate change and other environmental crises. For these authors, the state remains the essential political unit through which effective action needs to take place, albeit in the context of international regimes and agreements. These authors argue that the state possesses both the financial and coercive means, as well as the political legitimacy necessary to create the context for effective action on climate change. In short, the state, however imperfect, still provides a powerful and legitimate presence in world affairs when confronted by the increased power of transnational corporations in these globalized and neo-liberal economic times (Meadowcroft 2007).

One of the dominant ways the state has responded to climate change – particularly in the global North – has been to couch their responses within sustainable development discourses (SD). As was discussed in Chapter 2, as the modernist project was able to incorporate environmentalism (forming ecological modernization), development interests – previously seen as the anti-thesis of green concerns – were equally able to co-opt environmental opposition to the extremes of capitalism. Sustainable development was born.

SD can be understood in economic terms as attempts to resolve the traditional tension in environmental and later climate politics between striving for economic growth and protecting the environment (see Meadows et al. 1972). Instead of seeking to replace capitalism with some other alternative system of socio-economic organization, advocates of SD and EM suggest that governments, corporations and civil society can seek to promote economic growth, but that they must take greater responsibility for protecting the global environment from further damage. The Brundtland Report (World Commission on Environment and Development 1987) supported this interpretation of sustainable development by arguing that continued economic growth could support environmental protection, as well as promote social development (Doyle 1998). EM is typically interpreted as taking a variety of forms from weak EM (low levels of reform to the state with limited changes to prevailing models of political economy) to strong EM (entails a fundamental reconsideration of the structure of the state to open up policy processes to greater citizen participation and a fundamental reorientation of capitalism towards an ecologically-sensitive form) (Christoff 1996; Barry and Paterson 2004).

EM and SD have been an influential liberal discourse utilized in UK environment policy over the past two decades. Both Revell (2005), and Barry and Paterson (2004), argue that while there has been a visible shift in, for example, the UK's position towards EM, it is a weak version of it. Politicians from both Conservative and Labour parties have utilized the language of EM in their speeches on the climate and the environment. This is clearly understandable that at a time of increased ecological awareness, the 'win-win' (Revell 2005) philosophy of EM enables politicians to offer policy solutions to these threats that also contribute to economic growth. The influence of EM over UK environmental policy can be seen in the increased use of market-based policies. For example, the Climate Change Levy (CCL) was introduced in April 2001 as a tax on energy delivered to non-domestic users. The overall aim of CCL, which is applied to electricity, gas, coal and liquid petroleum gas for non-domestic users, was to increase energy efficiency and reduce carbon emissions.

Outside of Europe – but still in the global North – the whole debate pertaining to a 'carbon tax' was a defining moment in the lead-up to the 2013 Federal Election in Australia. The leader of the ruling Labor Party under then Prime Minister Julia Gillard was ousted by Kevin Rudd (who had been removed earlier by Gillard and her supporters) due, in large part, to a climate 'tax' policy which had proved immensely unpopular with the electorate, both within and outside the Party. In due course, rather than denying the existence of climate change as an issue (as had been the earlier response of the conservatives), the Opposition was able to chip away at Labor's support base by advocating 'market-led' approaches to the climate crisis, rather than the 'Big Government' approach of a carbon *tax*. As a consequence, the Abbott-led conservative opposition was able to win government on this, and other important policy stances.

This provides for us a segue way into the next section describing the ideological attributes of neo-liberalism and its advocates' economic responses to climate change. For this approach of the Abbott Government in Australia, like the current response of the Obama administration in the United States, no longer *denies* climate change; but rather, the state largely washes its hands of its climate responsibilities, handing over the survival of the planet to decision-makers in large corporations.

Climate chance and the economics of neo-liberalism

According to the economics of neo-liberalism, the market, not the state, lies at the centre of economic life. State intervention will usually produce sub-optimal outcomes (O'Brien and Williams 2010). Obviously neo-liberalism is a still a strand of liberalism. But as far as climate change is concerned, it is the dominant economic ideology driving policies both for mitigation and adaptation.

In the global North, climate change politics and economics are largely a battle between liberal and neo-liberal approaches, with the latter usually gaining ascendency. In the global South, as we will discuss in Chapter 7, debates pertaining to the economics of climate change also include the broader discussions derived from outside the political Right.

Radical libertarianism (or neo-liberalism), the third ideology discussed here housed in The Right, is still a form of liberalism, but is a particularly extreme, simplistic and, currently, virulent strand. Much of the radical libertarian/neo-liberal or 'market liberal' response is extremely reminiscent of laissez-faire economic systems which emerged in Europe last century coinciding with the industrial revolution (Doyle 2001). These economic libertarians advocate their own interpretation of Adam Smith's proposition that within capitalism existed *an invisible hand*, whereby the pursuit of individual self-interest leads, unwittingly, to the advancement of the common good. In this way, they seek maximum autonomy for the individual, and regard any government interference as a direct threat to this freedom and autonomy. It is rather obvious what 'the state of nature' is to radical libertarians: the survival of the fittest individual leads to the survival of the fittest collective, though the *condition* of the collective is hardly important. This is remarkably similar to popular and often misguided interpretations of Charles Darwin's theories of 'natural selection' and Herbert Spencer's theories of social evolution. It is hardly surprising that both emerged on the coat-tails of the industrial revolution and the initial surge of capitalism. This most extreme version of liberalism is really an anti-society ideology. Society is nothing more than a collective of atomized individuals; it is not designed to promote infrastructures of welfare, health and education let alone protect and uphold a collective sense of 'environmental good'. In the tradition of laissez-faire, the poor and the sick simply deserve to be so, and the 'management' of nature is best left to market forces, as the market *is* natural.

Green radical libertarian approaches have also been renamed free-market environmentalism. Attempts to make the *national accounts* and *consumerism* more greenhouse friendly provide excellent examples of this approach (Eckersley 1993: 29 – 30). In the first instance, green free-marketeers argue that a key problem is that not all of nature has been allocated economic value. Robyn Eckersley comments (1993:29): 'Many natural resources are regarded as gifts of nature with a zero supply price, and the accumulation of wastes is regarded as a 'negative externality'. For example, our woodchips exports are registered as national income but no allowance is made for 'natural capital depreciation' (i.e. the depletion of our forests).'

So rather than arguing that there are some natural attributes which are beyond fiscal value, green radical libertarians seek to include all of nature into their market strategies in a bid to resolve environmental problems whilst still pursuing economic growth under capitalism. Interestingly, while libertarians insist on the user-pays principle elsewhere, in discussions pertaining to anthropogenic climate change, they usually reject the *polluter-pays* principle.

Green consumerism, as touched upon in the previous section, is another central plank of all liberal green responses, but it is most avidly championed by the radical libertarians. Because of liberalism's associated commitment to the role of the individual, much emphasis is placed on the environmentally educated consumer. The argument goes: 'Once people start purchasing more climate "protective" products, then the market will have to respond and produce more of the same.' In addition, with so much emphasis on consumption, there is the notion that producers will self-regulate. This type of market-tactic is often aimed at 'empowering consumers', and is designed at reaffirming the importance of local markets. In Hungary, for example, the National Society of Conservationists (NCS) focuses upon 'conscious consuming', 'buying Hungarian', with the slogan of 'Your purchase is a vote' (National Society of Conservationists Hungary 2010). The NCS lists six tactical tips aimed at the individual consumer/citizen who is both climate change and environmentally aware:

1. Conscious consuming.
   Consume circumspectly, look thoroughly, what can be found in the product, where it comes from and how was it grown or produced.
2. Your purchase is a vote.
   During shopping you vote for a given production system (home, organic/import, intensive) assisting with the money you spent.
3. Buying home products
   If you search for home farmer's products, your money will stay inland so you can contribute conserving home workplaces. It's very important also that you can reduce the pollution of transporting, if you don't buy import products.
4. Buying organic products.
   Consuming organic products you support a system which make high quality, healthy products that are free from pesticide residues.

Because they don't use chemical fertilizers and other chemicals during cultivation you protect the environment as well.

5. Buying seasonal vegetables and fruits.
   You shouldn't eat glasshouse vegetables and fruits from another part of the world but the fresh, currently ripened. The pollution of transporting will reduce and you don't support the intensive, industrialized cultivation.

The solution, however, is still firmly entrenched in the politics of pluralism and capitalism which assumes that producers exclusively respond to markets, rather than setting and influencing marketing agendas.

This 'climate co-option' repositioning by the ultra-right now accepts the veracity of anthropogenic climate change but remains firmly against the liberal, more co-operative and state-centric Kyoto Protocol as the 'right way' of addressing the problem (instead advocating voluntary business partnerships). The essence of this rhetoric first emerged forcefully in the public realm during the inaugural meeting of the Asia-Pacific Partnership on Clean Development and Climate in Australia in January 2008. As well as bringing together the two non-compliant nations – Australia and the US – to the climate change table, it also managed to secure the participation of China, India, Japan and Korea. The major theme of the conference was that business, not nation-states or Kyoto, was the salvation to global climate change. *The Weekend Australian* reported (14–15 January 2008: 16) in an opinion piece without byline, continued with its own ideological rant along these lines:

> The reactionary response to the Asia-Pacific Partnership meeting this week demonstrates that support for Kyoto cloaks the green movement's real desire – to see capitalism stop succeeding. Extreme greens cannot bear to accept that our best chance of reducing greenhouse gas emissions will occur when free enterprise has incentives to implement solutions. (*The Weekend Australian*, 14–15 January 2008: 16)

So, in this vein, 'extreme greens' are created who construct Kyoto as a communist model (when it is in fact simply a liberal co-operative model of global economic governance described in the prior section), whilst these right-wing, neo-liberal think tanks paint themselves as moderate greens, now accepting the current climate crisis, but offering a different solution: a roundtable built by business, not the nation-state. In this view of geopolitics and geoeconomics, climate change will be resolved by the very perpetrators of the crisis out of some sense of corporate moral and social responsibility. In fact, this model is so attractive to big business industries, such as transitional extractive industries, as it will lead not quite to better-than-usual, but to business better-than-usual. Cate Faehrmann of the New South Wales Nature Conservation Council argues that the APPCDC's six nations voluntary approach was a 'license for government and business to do nothing ... Without any incentives or penalties there is no reason for industry to move away from burning coal and fuel' (Faehrmann quoted in Wikinews, 12 January 2006).

In this situation, a notion of a liberal 'carbon tax' is replaced by a free market 'cap and trade' solution. Cap-and-trade systems need long-term emissions caps that place 'an unambiguous limit on the amount of carbon dioxide permitted to be released into the atmosphere over the long haul' (Kurtzman 2009). Unfortunately, ambiguity seems to be the only current factor which is certain over the long haul when corporations are largely governing their own programs (within limits set into legislation by a partnership between the state and big business). Kurtzman writes of faith in market climate mechanisms in the context of the United States. A good example lies with the American Clean Energy and Security Act of 2009, a response to the United States' failure to ratify the Kyoto Protocol in 1998 (ratified by 183 parties, including all the developed countries except the United States). The bill granted new authority to the Environmental Protection Agency (EPA), the Commodity Futures Trading Commission, and the Federal Energy Regulatory Commission, and also established a registry of greenhouse gas emissions. It also included an attempt at measuring efforts to offset emissions, 'such as planting trees, transforming animal waste into methane gas for energy use, and capturing methane as it escapes from landfills'. But, most importantly, the key component of this legislation was the free market, carbon-trading response. He writes:

> The goal is to gradually reduce U.S. greenhouse gas emissions to 17 per cent of 2005 levels by 2050, beginning with a modest three per cent reduction by 2012. The bill would require reductions in emissions from most stationary sources of greenhouse gases, including power plants, producers and importers of industrial gas and fuel, and many other sources of carbon dioxide, such as steel mills and cement plants. It would also raise mileage standards and lower permissible emission levels for vehicles. Crucially, the bill puts its faith in the *market* and its ability to lower the cost of reducing emissions through the trading of permits. (Kutzman 2009: 114–122)

This is not, then, an ideological battle between socialism and capitalism, as the aforementioned 'opinion piece' above suggests, but one between two ideological strands of the right: liberalism and neo-liberalism. Like neo-liberals, the classical liberal position argues for the primacy of the individual, and upholds the basic tenets of the capitalist economic system. But liberalism differs from its neo-liberal or radical libertarian cousins, as it maintains that the state has a role to intervene in human affairs when the will of the anti-social minority (in this case environmentally degrading companies), interferes with the wishes of the majority. In this light, the Kyoto Protocol is firmly entrenched in the liberal tradition, as it sees a role for responsible, democratic governments coming together to provide sticks as well as carrots to alter poor, climate-degrading, national and corporate practices.

The neo-liberals (intriguingly informed by a conservative moral position), on the other hand, see any state intervention in the marketplace as anathema. Rather than championing democracy, the neo-liberals actually champion plutocracy: the wealthy should decide. Further, the earth itself is simply a large corporation, and nature is the market – for the market is deemed natural, per se. All climate and other environmental issues will resolve themselves if only left to the free hand of the market principles. In the North, in both Australia and the United States, this current ideology remains rampant, whereas in Europe – the champions of Kyoto – a less virulent strand of liberalism combined with conservatism is dominant. But even ex-Prime Minister Tony Blair's 'Third Way' can hardly be interpreted as a socialist conspiracy to overthrow capitalism. In this manner, there is much discussion about 'climate sustainability': about giving ticks to industrial practices which partly reduce emissions, and to support businesses in emissions-trading. Finally, there is a smattering of discussion about 'climate footprints', giving a nod to less carbon-based forms of energy, promoting, for example, kinder forms of air-conditioning, as well as developing passive solar housing etc.

Risk and climate insurance

Another key way in which the climate change debate has been marketized in neo-liberal terms has been through the politics of *risk* and, as a consequence, the engagement of the insurance industry. The language of risk first emerged in policy-making circles from *within* corporations. Cordner writes:

> An internationally accepted risk vocabulary has been developed over many years of application, primarily in the practical world of private

> enterprise. *ISO Guide 73* (ISO 2009: 1–2) defines *risk* simply as the 'effect of uncertainty on objectives'. (Cordner 2014)

The use of risk as an overarching 'ordering principle', first by markets, and now implemented by states, is the result of, Ulrich Beck would argue, fundamental changes in advanced capitalist societies. Beck refers to this new polity and economy as 'reflexive modernization': 'the age of uncertainty and ambivalence, which combines the constant threat of disasters on an entirely new scale with the possibility and necessity to reinvent our political institutions and invent new ways of conducting politics' (Beck 1999: 93). Obviously the climate crisis fits in neatly here as a new kind of transnational disaster. Beck went on to argue that the 'very idea of controllability, certainty or security' disappears, and a 'world risk society' emerges where all nations and cultures share the same risks.

Obviously, as argued previously in this book, and again taken up in Chapter 6, this concept of risk has profound implications on constructions of security, and the divisions between economics and security often fade away. But for the purposes of this chapter, it is enough to argue that such a concept, derived from markets, imagines all of the planet (and its climate) as a market. The insurance industry has seized upon this concept of risk as a means of not just addressing the climate crisis (and, it is argued, addressing *vulnerabilities*) but also as a means of maximizing profitability within existing Northern markets, and entering new marketplaces in the global South. Insurance used to be a luxury position of the North (it was an option which could be taken); but with climate change as the transboundary stick, insurance no longer is construed as optional. The least affluent are forced within its rubric; non-complying cultural behaviors are challenged; populations are disciplined by market; values relating to risk with 'halo' (western) behaviors rewarded.

Insurance companies have been swift to acknowledge the costs of climate change to their industry. Mills (2009) argues that by 2009, the insurance industry was clearly aware of the threat to its own business. Its own modeling echoed the results of climate scientists (ibid: 325), and insurers themselves were naming climate change as one of their principle challenges. Mills cites a 2007 PricewaterhouseCoopers survey of 100 insurance industry representatives who put climate change as their 4th issue out of 33, and a 2008 Ernst and Young survey of more than 70 insurance companies which rated climate change as their number one risk (ibid.: 324). Some insurers formed interest groups to deal with the issues such as ClimateWise and the UNEP Finance Initiative.

The industry now recognizes that climate change adaption and mitigation will itself bring new risks (increased use of nuclear energy, or emerging green technologies for example), but it also sees climate change as an opportunity.

The market is already reacting. Shareholder resolutions regarding climate change hit a record in 2008 (ibid.: 328). Mills argues that there will be increased demand for conventional insurance, but also sees opportunities for new products like *weather derivatives and catastrophe bonds*, and so-called 'innovative products' like *micro-insurance and public-private partnerships*, which he thinks 'will allow markets to grow and serve the billions of people in the developing world who currently lack insurance' (ibid.: 329).

As touched upon, much of the global South has traditionally fallen outside the insurance industries market reach and climate crisis provides it with a unique opportunity to further extend its market reach into the global South. To do this, they are also collaborating with a host of non-insurance groups, such as energy utilities, foundations, or governmental agencies: 'Recent examples include the Earth Institute at Columbia University working with Swiss Re to implement satellite-based remote sensing in support of micro-insurance for small farmers in Africa and a joint project between Munich Re and the London School of Economics to refine our understanding of the economics of climate change' (ibid.: 335). Mills writes of this new global reach and the promise of these emerging micro-insurance markets in the lands of the poorest. ClimateWise, launched in 2007, speaks of the importance of increasing entry into countries which possess 'low insurance penetration':

> Climate-related micro-insurance, which provides coverage for low-income populations without access to traditional insurance, is reaching a greater number of policy-holders than most climate-related products in the traditional market. This paper identifies micro-insurance products covering about 7 million policy-holders. Many of these products respond to climate-linked vulnerabilities such as food and water shortages in rural areas of South America, Africa and Asia; much of this market activity tends to be driven by European insurers. (Mills: 314)

Veron and Majumdar (2011: 31–41) critically examine micro-insurance (including weather-based insurance), through corporate-NGO partnerships in West Bengal. In this context, NGOs are heavily involved in getting individuals to participate in the nascent micro-insurance products (regulated but also promoted by the Indian Government after its privatization of previously state-based schemes). The difficulties explained here have a local and national political context: micro insurance is likely to have different results, and problems, in different places. Local NGOs are recruited by the companies in West Bengal, as it is they who have the contacts and networks with local communities. These NGOs, at least in the early days of engagement with the insurance companies, are relatively happy to participate because they think it will reduce their reliance (and the local peoples') on donors. The problems found were: companies became frustrated because they were using traditional insurance approaches and products which did not work; NGOs were similarly frustrated by the insurance companies' heavy-handed approach to 'capturing' customers to satisfy business targets; and villagers became disillusioned as they simply struggled to understand the concept of *western* insurance at all (they were rather more used to 'risk-pooling' in the form of '*balanced reciprocity*, where some return can be expected at some point in time, and not *mutual insurance*, where only the unlucky are compensated' (Veron and Majumdar 2011: 127)). There were also local political dynamics involving the local left-wing government which further complicated the panorama.

Luis Lobo-Guerrero writes of this manner in which the poorest people are incorporated into the economics of risk. He talks of the ways in which 'the state's traditional sovereign responsibility to protect' is transformed 'into a responsible risk management enterprise'. In this way the markets of the global South are accessed as never before, 'transforming the uninsured livelihoods of people (*the global South – authors' addition*) into insured lifestyles of populations' (*the global North – authors' addition*) (Lobo-Guerrero 2011: 90–91). Whether the third world remains interested in climate insurance relates to whether or not it can receive compensation for past climate-induced weather-events; but affluent world insurers are only interested in paying out for future events. There are also serious concerns that as corporate climate responses engage with village communities, there is a disciplining of these communities to re-organize outside traditional, social and cultural lines. In this manner, local peoples are disciplined to conform to the rigors of northern-markets in an economic game in which they simply cannot win.

On other occasions, regardless of any normative gains from companies assuming a 'moral' or 'socially responsible position' in relation to climate change, extreme weather events interpreted through the lens of climate change offer an easier form of 'event auditing' and, ultimately, most development in relation to climate insurance has been pointed firmly and squarely to the advantage of the affluent world. As a consequence, there has also been a huge surge in business as the insurance industry seeks to protect the wealthiest, resulting in a boom in liability insurance:

> Almost all of the climate-related innovations in liability insurance for directors and officers, political risk, professional liability and environmental liability have appeared in the past year. Both Zurich and Liberty Mutual launched products specifically designed to cover boards of directors in the event of climate change litigation, a significant development given pending lawsuits that could allocate significant costs to major emitters of greenhouse gases. (Mills 2009: 340)

To conclude this section, it must be understood that much of this wedding of climate change with insurance has little to do with the industry itself, but with nation-states. Nation-states, once willing to be the core provider of essential services to the Earth's citizens, now also operate along corporate, neo-liberal principles. Their prime interest is no longer in protecting the planet and its peoples, but in externalizing and defraying climate risk (Blazey and Goving 2007) in a manner which protects the state and corporate elites. The socialization of risk, using governments to 'share the burden' of anthropogenic climate change, is a key motivating factor behind this post-political and neo-liberal market response to such environmental crises.

Conclusion

Climate change works hand in hand with free market economics. As the market is deemed 'natural', the ecology of the ecosphere becomes 'the market'. All inputs and outputs are given value in monetary terms and then, so it is argued, the 'natural', 'real' and 'essentialist' economy of ecology shall emerge, unfettered by the constraints of science and governmentality. It is imagined that 'the trickle down effect' will benefit those species living on the lower rungs of the natural hierarchy, promoting widespread ecological health and forever doing away with any notions of science-generated ecological safety nets thrown over the most disadvantaged, those species and habitats, those resources, most at risk. Through carbon-trading, nothing is irreversible; everyone and everything will win: the chocolate ration has been increased. This line of argument, of course, also fits neatly into the parameters of neo-Spencerism, promoting the notion that those human and non-human communities most likely to survive unfettered 'natural' systems will be all the better for doing so, having weeded out those less able to survive.

Of course, there are other responses to climate change which are not neo-liberal, or even from the ideological Right (see Chapter 7). But, no doubt, it is the neo-liberal market responses which are most dominant. Carbon trading, as aforesaid, is the most obvious form of the neo-liberal response, as it promises to deliver us to a carbon-neutral future by trading between the affluent world and the less affluent world. In this vein, the global South sells it future capacities to produce carbon (in effect, to industrialize) to the North which not only continues to produce emissions, but, in real terms, increases them. Importantly, as touched upon in the previous chapter, carbon-trading and other neo-liberal solutions can only succeed if the very nature of the liberal subject is challenged, and turned into a neo-liberal, 'post-political' subject: the global citizen-consumer.

With the global polity understood as a pluralist and post-political one, played out on a notional level playing-field, each citizen is actually a consumer. All people are considered 'equal' under this model; they just need *listening to,* or entry into the market-place. Above all, these post-political subjects need to be plucked out from the non-integrating, black spaces of the planet into the light of Northern markets, and one key way to do this is for all to be *insured.* There is no longer a clear delineation between *us* and *them*; haves and have-nots; subjects and objects of environmental degradation. All beings in this supposedly *post-political* realm are seen as consumers and providers, all meaning systems of collective action which may generate opposition are done away with because, again: 'We are all just people, just global citizens'. Questions of gender, class, and race – it is imagined – melt away.

It is proposed that the salvation of the earth from anthropogenic climate change will come from *within* individuals (by changing their behaviours), not in the politics of communities, or even in the politics of liberal nation-states. *Nature,* or in its current guise, *Climate,* is further commodified within this framework. Principles of sequential use and climate change, are almost totally committed to global values change, whilst denying the existence of any power differentials between cultures and people. As mentioned previously in this book, this is akin to the construction of a *global soul.*

Furthermore, the unwritten text of this process is that there is something wrong with the values and behaviors of the *majority world*. Environmental problems, like climate, are seen as getting worse due to the incorrect value systems of the South, rather than from the pressures on local communities to remain competitive in the new globalized, free market-place. The reality is that climate economics is largely a Northern document, a map with clear lines of cultural imperialism. Of course, the dissolution of the concept of *the Other* has not occurred in the South: in fact, the boundaries have often solidified with the acknowledgment that globalization has delivered disproportionate amounts of wealth and power to Northern and Southern elites (Doyle 2005).

By constructing all of the Earth's citizens as one amorphous mass, *incorrect assumptions about the South* lead to additional pro-northern biases. Modeling for climate change abatement models ignores economic implications for the South by making 'blatantly incorrect assumptions regarding the structure and dynamics of lesser-industrialized country economies' (Kandlikar and Sagar 1999: 130). These models ignore informal economies, market disequilibriums, and the different developmental choices which necessarily affect investment patterns, the nature of institutions, markets and competition, and the nature of market information (ibid.: 131–132). What is more, the climate change debate is not framed historically, but only in a future sense (likely future emissions), thus absolving the major emitters of responsibility with the issue of equity neatly being further ignored.

This leads to a crucial point: in most parts of the world, for many years, *climate* was seen as a non-issue – an issue constructed by western science, and then utilized as an environmental security issue to control the less affluent from pursuing the very path of development which the minority world has pursued without restraint since the scientific and industrial revolutions. To many in the majority world (where most of the people live), the key environmental economic issues have nothing to do with barely perceptible changes to atmospheric temperatures. Instead, they revolve around providing immediate and secure access to resources which provide people with the basic components for survival: nutrition, shelter, freedom from disease, a sustainable and just livelihood, and safety (Doyle and Risely 2008). Both the writers of this book have worked for a long time with the Indian Ocean Research Group (IORG) working with countries and communities across the 'Ocean of the South' (Chaturvedi 1998), and know of the most critical environmental/economic issues in the poorest parts of the world first hand: people displaced and dying over water, air, food and energy conflicts. Yes, climate change may contribute to these conflicts; but far more importantly, those who initially survive these shortages and conflicts often become environmental refugees and migrants, on a scale of millions unimaginable, often fleeing as a consequence of human intervention in the short-term – as a direct consequence of large multi-national companies working in close association with national elites, pursuing massive extractive operations with enormous environmental costs in their countries. Millions, for example, have been displaced by the building of big dams which are currently spreading across the third world, interconnecting entire river systems. Millions are displaced by agro-industries, planting vast areas of monocultures, destroying the very fabric of the land and communities by taking away ancient systems of sustainable and mixed farming, and then patenting seed-stocks, giving them a 'terminator' gene, so that people cannot use them again. Millions are displaced by forestry companies clear-felling vast areas of remaining forests, and displacing even more by planting out existing agricultural land with fast-growing species such as palm oil trees, blue gum and pinus radiatia, and then receiving carbon-credits for being good 'climate change' citizens. Millions are displaced by large mining interests building open cut mines and then polluting entire river and marine systems with their partially and imperfectly treated waste products. Millions are displaced by transnational fishing companies who control and deal in what the United Nations calls the most traded commodity on the planet. Most environmental security problems in the majority world are *economic*, they are about the market, but in a very different way that discussion around *carbon markets* normally entail. Instead, they are about unequal access for local people to their resources; and they are about maldistribution of those precious resources, where most money made from these short-term enterprises goes into the pockets of local elites and/or back offshore.

In short: Laissez-faire ecology is a nonsense and non-science. It does not acknowledge the widely recognized trend away from Darwinist competition models in biology. Instead, it nicely supplements the (sometimes post-) politics of neo-liberalism and the globalization of advanced capitalism. Within it, there is no understanding of the reality of relationships within and between societies and ecosystems. By imagining nature (or *the climate*) exclusively as an ecological free-market, it will lead to massive species extinction and habitat degradation, whilst also extending inequities between human beings. Unrestrained commodification of the planet Earth will not lead to wise use, win-win games, but will, in the short-term, lead to very few, but extremely large wins for a powerful minority, and in the long-term, it will lead to massive losses for all forms of life on the planet.

5
'Climate Borders' in the Anthropocene: Securitizing Displacements, Migration and Refugees

Introduction

At the heart of contemporary, fast multiplying climate-security narratives originating largely (but not entirely) from the global North are the imaginative geographies of millions of impoverished Afro-Asians being uprooted and displaced from their habitat and crossing borders in search of the greener and securer pastures. The latest 5th Assessment Report of the Intergovernmental Panel on Climate Change (IPCC) has for the first time added a detailed discussion on 'human security' and 'sustainable development' in Chapter 20 of Working Group II. This undoubtedly is a welcome addition to the IPCC agenda and to some extent blunts the social science critiques of its earlier four assessment reports. Having said that, the citation above raises a number of complex questions, largely unaddressed, about the *context* in which to approach the complex but connected issues of climate induced displacements (cited hereafter as CID) and climate induced migrations (cited hereafter as CIM).

Once the issues referred to in this citation (i.e. poverty, violence, economic inequalities, loss of biodiversity etc.) are approached as interconnected outcomes of what John Bellamy Foster (2009), inspired by Marxian thinking, would refer to as 'metabolic rift', a far more systemic perspective emerges. In this perspective (which we prefer for the purposes of this chapter) political ecology and political economy are two sides of the same coin that is capital accumulation; 'the necessity of continued, rapid growth in the production and profits' (Foster 2009: 57). The notion of the *metabolic rift* suggests that,

> the logic of capital accumulation inexorably creates a rift between society and nature, severing basic processes of natural reproduction.

> This raises the issue of ecological sustainability – not simply in relation to the scale of the economy, but also, and even more importantly, in the form and intensity of the interaction between nature and society under capitalism. (Foster 2009: 49)

As imaginative geographies of threatening incremental climate change seek embodiment in the figure of a 'climate migrant', new walls/borders of otherness (both mental and physical) are being erected and in the process the old ones are reinforced. Who and how many of these millions (terrorizing figures indeed) will be migrating from one place to another in search of more secure spaces within their respective national boundaries and how many would dare to cross international borders? The answers are not easily forthcoming. And yet this is *the* category of CIM (which encompasses the notion of 'climate refugees') that seems to have caught the most strategic attention and imagination of fast multiplying 'national security' and even 'human security' narratives.

Our central argument in this chapter is that similar to the 'war on terror', the so-called 'war on climate' invokes – through the deployment of certain metaphors – a borderless, flat 'global society' at 'risk', but the practices it gives rise to are resulting in highly territorializing but invisible borders both within and across national boundaries. These new fences and walls (both material and discursive) are being conceived, constructed and imposed by the 'minority world' in anticipation of a large number of 'climate migrants' fleeing from the 'majority world'; an overwhelmingly impoverished world that is allegedly falling terribly short of 'capacity' and 'resilience' while struggling to adapt to ever soaring temperatures.

Ironically enough, some of the imaginative geographies of a 'planetary emergency', demanding a universal-global governance response, construct (through a techno-market approach), inward looking 'climate territories' that in all probability might terrorize millions of displaced populations by frightening them away from asking value loaded questions of political accountability. These marginalized displaced communities now face the specter of being re-located (doubly displaced) into new contexts and texts of the 'Anthropogenic' and the 'Anthropocene', where 'expertise' is being privileged over day-to-day experiences of livelihood struggles on social-political margins. It is the politics of the 'non-political', 'apolitical', and 'post-political' discussed by us at some length in Chapter 1, that is likely to subject the losers of neo-liberal globalization to 'climate terror', as the market-military combine gains salience, bordering hegemony, in climate change discourses.

For the purposes of this chapter we restrict the illustrative part of our analysis largely to Bangladesh. The reason we choose to do so is two-fold. First, wary of generalizing the bordering cause-effect of the CIM across varying scalar, spatial and power-political equations, we believe it is important to acknowledge the role played by the geo-historical and geopolitical specificities. Our choice of Bangladesh we hope, aptly illustrates the point we are trying to drive home in this chapter. The geopolitics of 'climate borders' is entangled in growing contestations between impulses of national security and imperatives of human-livelihood security. Second, we would like to argue and illustrate that Bangladesh is a good example to show how climate change discourses, despite their boundary defying, global commons rhetoric, are causing the 'thickening' of already existing stubborn borders.

This chapter is based on the premise that the science of climate change is fast becoming a powerful orthodoxy amongst many intellectuals, governments, corporations and non-governmental organizations, particularly in the global North. In recognizing this dominant category in scholarly and political discourses, our key intention here is not to deny or validate the premises and conclusions of climate change scientists in any essentialist manner, but to build on and develop the insights offered by a number of recent studies by political geographers exploring the how and why of the discursive production of geographical knowledge (in plural) of climate change by various actors/agencies, in support of certain domestic as well as foreign policy agendas. We argue that it is the geopolitics of fear that appears to be dictating and driving the dominant climate change discourses in Bangladesh. The chapter first develops a theoretical perspective through which to analyze the imaginative geographies of climate change-induced displacements and their implications for South Asia. Next, we focus on various facets of the geopolitics of fear and on some of the key sites where climate change knowledge production about Bangladesh is taking place. One of the ways in which climate change is folded into a discourse of fear (that, in turn, requires a geopolitical response) is by referencing the 'problem' of refugees. Penultimately, then, we then move on to deconstruct the official discourses and political speeches both within Bangladesh and its immediate neighborhood in India, in order to reveal the underlying geo-politics of fear and boundary-reinforcing cartographic anxieties about climate change-induced displacements and migrations. We conclude the chapter by examining the prospects for counter-imaginative geographies of hope and the role they could possibly play in approaching the issue of CIM from the angle of human security and human rights of the socially disadvantaged, dispossessed and displaced in the global South.

The invisible world of the 'Displaced': perspectives *on* and *from* the Global South

The International Organization for Migration, while mapping out the complexity surrounding the issue of climate-induced displacements and migrations, has been sounding a note of caution against the geopolitics of exaggeration from time to time (IOM 2009). A major catalyst for research on issues related to migration and environment appears to be the 'fears that millions of people from some of the poorest countries in the world could be forced to migrate to richer parts of the world due to climate change.' In order to ensure that the research agenda on migrations is not framed too narrowly, 'It is essential to start from the position that migration is not always the problem, but can in certain circumstances, where migration contributes to adaptation, be part of the solution. In short, migration linked to climate change will create both risks and opportunities.' What is needed therefore is far more serious and systematic research aimed at teasing out the complex relationship between the environmental/climate change and migration. It is not inevitable that environmental and climate change 'will in all circumstances automatically result in the increased movement of people.' For example, the 'number of persons affected by natural disasters has more than doubled in recent years but we have not seen a major increase in international migration in many of the disaster affected regions.'

Even a cursory glance through the website of Internal Displacement Monitoring Centre (IDMC 2013) reveals that highly alarmist accounts of climate induced migrations make the universe of those 'frightened and forced away from their homes by armed conflicts, communal violence, abuses of human rights and humanitarian law, natural or man-made disasters' more or less invisible. A vast majority of victims simply do not have the capacity and resources to cross international borders and maybe that is one of the reasons for their relative invisibility. 'Whereas in 1982, it was estimated that some 1.2 million were forcibly displaced in 11 countries, by 1995 an estimated 20 and 25 million IDPs were located in some 40 countries, approximately double the number of refugees worldwide' (Goldman 2009: 38).

According to the Internal Displacement Monitoring Centre (IDMC 2013: 6), during 2012, disasters associated with natural hazard events added nearly 32.4 million people in 82 countries to the category of displaced. And out of around 144 million people forced from their homes in 125 countries, between 2008 and 2012, nearly three-quarters were affected by multiple disaster-induced displacement events. Not surprisingly, as a result of repeated displacements, the overall resilience of the victims was seriously impaired and the affected communities were caught up in an unending vicious cycle of multiple vulnerabilities.

It is important to note that the figures cited above relate to just one category of internal displacement caused by natural hazard events. According to yet another report by the same organization (IDMC 2014: 9),

> There were 33.3 million internally displaced people in the world as of the end of 2013. They were forced to flee their homes by armed conflict, generalized violence and human rights violations. This figure represents a 16 per cent increase compared with 2012, when we reported 28.8 million IDPs, and is a record high for the second year running.

According to this report *at least* 526,000 in India, 631,000 in Afghanistan, 746,700 in Pakistan, 280,000 in Bangladesh and 90,000 in Sri Lanka were displaced under this category. And in the case of Palestine the number is said to be *at least* 146,000 (ibid.).

Geopolitics of 'climate borders': 'destabilizing orders' and 'threatening others'

David Newman (2010) argues that 'Borders are the constructs which give shape to the ordering of society ... Notions of territory and borders thus go hand in hand.' There is much more to a border than its physicality and materiality on the ground. Borders are conceived, constructed, imposed and even resisted primarily through emotional geographies that are far from being politically innocent or eternally disembodied. These boundary producing (between 'us' and 'them') imaginative-emotional geographies are often deployed at the service of power-political-policing practices of the institutions of statecraft.

As pointed out in Chapter 1, a critical geopolitics of climate change, in our view, enables us to expose various scripts and narratives of climate change in terms of a knowledge power nexus, and to explore how they frame various places on selectively drawn regional and global maps of threats and insecurities. What provides an extraordinary complexity to such scripts and their imaginative geographies of fear is that all of them, despite political ideological agendas of their own, claim to derive their respective authority, legitimacy and efficacy from the natural science evidence of global warming and climate change; a point to which we shall return shortly in the section to follow.

Dominique Moïsi in his book entitled *The Geopolitics of Emotion: How Cultures of Fear, Humiliations and Hope are Reshaping the World* (2009) makes a number of thought-provoking observations. His argument is that geopolitics is not only about materiality and resources but also about emotions; 'one cannot comprehend the world in which we live without examining the emotions that help to shape it' (Moïsi 2009: xi). According to him, the reason that he has chosen 'fear', 'hope' and 'humiliation' as the key emotions for analyzing contemporary global geopolitics is because all three are related in one way or another to the notion of confidence: 'which is the defining factor in how nations and people address the challenges they face [e.g. climate change] as well as how they relate to one another' (Moïsi 2009: 5). According to Moïsi,

> Fear is the absence of confidence. If your life is dominated by fear, you are apprehensive about the present and expect the future to become more dangerous. Hope, by contrast, is an expression of confidence; it is based on the conviction that today is better than yesterday and that tomorrow will be better than today. And humiliation is the injured confidence of those who have lost hope in the future; your lack of hope is the fault of others, who have treated you badly in the past (ibid.).

Moïsi's argument is that after having dominated the Westphalia state system for almost two centuries, the West is now in the grip of acute cartographic anxiety, feeling increasingly vulnerable and insecure due to perceived loss of control over fast multiplying forces emanating from the global South, including immigration. In his view, in its most dominant variant, 'fear is an emotional response to the perception, real or imagined, of an impending danger' (Moïsi 2009: 92). Moreover:

> In the last few years a new cycle of fear, one that shares many common features in Europe and the United States, has invaded our consciousness. I do not think it actually began with 9/11, which only confirmed and deepened it. In both regions of the West, this new cycle includes fear of the Other, the outsider who is coming to

> invade the homeland, threaten our identity, and steal our jobs. In both regions, it includes fear of terrorism and fear of weapons of mass destruction, the two being easily linked. It includes fear of economic uncertainty or collapse. It includes fear of natural, environmental, and organic disasters, from global warming to disease pandemics. In sum, it involves fear of an uncertain and menacing future, over which there is little, if any, possible human control. (ibid.: 94)

It may be that the writings of Moïsi (however attractive to our purposes here) are, in part, problematic. His ascription of characteristics to countries and regions is often contrary to the emphasis on contingency and aversion from time-transcendent essences that characterize so much of our argument. Regardless of these foibles and partial logical inconsistencies, it can be said, however, that the geopolitics of fear is fundamentally conservative. It draws upon and, in turn, feeds into various alarmist imaginative geographies and sits too easily alongside realist schools of thought within the most hide-bound and archaic traditions of international relations scholarship. With world spheres constructed within this tradition as an anarchic system of mercantilist nation-states engaged in zero-sum political and economic games, the fear of chaos lies at the heart of Westphalian political dreaming. In turn, this fear of chaos is often used to justify the use of the big Northern 'stick of reason', to discipline and to order the global South to supply the much needed natural authority to control the imminent chaos (Doyle and Chaturvedi 2011), whether through agendas of climate change, military intervention, development, or modernization.

As we will argue and illustrate in the following sections dealing with Bangladesh, the written geographies of climate change science, imaginative geographies, therefore, legitimize and create 'worlds'. Paradoxically enough these imagined worlds are constructed among other things through simultaneous metaphorical invocation of a borderless, flat global space and a tamed minority world, (having unconsciously committed the 'crime' of polluting the atmosphere due to the then lack of scientific evidence since the industrial revolution), at the receiving end of consciously polluting fast growing economies of Asia.

In his thought-provoking book entitled *Security and Borders in a Warming World: Climate Change and Migration, Gregory White* (2011) draws attention to how the concept of 'climate-induced migration', despite considerable ambiguity and contestation surrounding it, has started galvanizing military-security-intelligence agencies around the world, especially in North Atlantic; a region he describes as including the US, Canada, the European Union (EU), and its member states. In our view, he is absolutely spot on in his observation that,

> 'Getting tough' – responding in a militarized fashion – is an easy, cynical step in a warming world. It may be politically successful with anxious electorates. It may tap into the public's fears about climate change and the prospect of desperate hordes of 'refugees' inundating North Atlantic borders. And it may be more politically palatable than policies that mitigate greenhouse gas (GHG) emissions. Building a fence is easier than changing lifestyles. Yet the injection of security imperatives into climate-induced migration is unethical and unworkable (White 2011:7)

White's critical gaze focuses on the case of Sahelian and Sub-Saharan African migration to Europe, and shows how the 'unethical' and the 'unworkable' are causing considerable cartographic anxieties among institutions of statecraft that remain somewhat embedded in highly resilient Westphalian territoriality even in the case of EU, the most celebrated example of the post-Westphalia experimentation with the traditional notions of sovereignty and security. We also share his concerns that what is needed is a careful mapping of the complex 'empirical realities that affect migration factors' (the so-called push and pull factors), a critical reflection on how to ''desecuritize' the discourses and policy associated with CIM', and while questioning the growing militarization of climate change, 'craft an ethical and practical set of policy initiatives to address climate change and the migratory flows to which it might contribute.'

Framing Bangladesh as a climate 'black hole'

In the vortex of alarmist imaginative geographies of 'catastrophic' anthropogenic climate change, Bangladesh is being increasingly implicated as a 'black hole'. These reports occur at multiple sites: in many think tanks engaged in strategic forecasts and planning (CNA Corporation 2007), official discourses and speeches, non-governmental organizations and the media. At the heart of this geopolitics of fear is the widely circulated image of a densely populated region. In 2006, 140 million people lived in an area of 144,000 km2 at a density of over 950 persons/km2. Furthermore, it is low-lying (two-thirds of the country is less than five metres below sea level) and natural disaster-prone (e.g. six severe floods in the last 25 years). This country's position in the Bay of Bengal has made it the source and site of millions of displaced and dispossessed 'climate migrants' and 'climate refugees'. Both the manner in which Bangladesh has come to embody the abstract notion of 'climate' (and 'dangerous climate change') against the backdrop of its long standing history of ecologically unsustainable 'development' and 'natural disasters', and the ways in which the so-called 'climate refugees' are being discursively transformed into unwanted, threatening internal and external 'Others', demands attention of a critical social science of climate change.

Scientific framings of climate change and their implications for Bangladesh: conceptualization and contestation

At the forefront of the 'scientific knowledge' production about climate change is the Intergovernmental Panel on Climate Change (IPCC), the mandate of which, according to the UN General Assembly, is to undertake international assessment of the current state and status of scientific knowledge about climate change, to examine its impacts and the range of possible mitigation and adaptation strategies. As a hybrid agency comprising scientists and bureaucrats 'it was to be governed by a Bureau consisting of selected government representatives thus ensuring that the Panel's work was clearly seen to be serving the needs of government and policy. The Panel was not to be a self-governing body of independent scientists.' In reality, as Hulme (2009: 96) points out, the boundary between science and policy is neither easy to maintain nor to 'police' (Hulme 2009: 96).

The IPCC Fifth Assessment Report (AR 5), in Chapter 5, entitled, *Climate Change 2014: Impacts, Adaptation, and Vulnerability*, focuses on 'coastal systems and low lying areas'. It is pointed out that,

> For the 21st century, the benefits of protecting against increased coastal flooding and land loss due to submergence and erosion at the global scale are larger than the social and economic costs of inaction *(high agreement, limited evidence). Without adaptation, hundreds of millions of people will be affected by coastal flooding and will be displaced due to land loss by year 2100; the majority of those affected are from East, Southeast and South Asia (high confidence)*; At the same time, protecting against flooding and erosion is considered economically rational for most developed coastlines in many countries under

> all socio-economic and sea level rise scenarios analyzed, including for the 21st century GSML rise of above 1 m (*high agreement, low evidence*).
>
> The relative costs of adaptation vary strongly between and within regions and countries for the 21st century (*high confidence*); *Some low- lying developing countries (e.g. Bangladesh, Vietnam) and small island states are expected to face very high impacts and associated annual damage and adaptation costs of several percentage points of GDP; Developing countries and small island states within the tropics dependent on coastal tourism will be impacted directly not only by future sea level rise and associated extremes but also by coral bleaching and ocean acidification and associated reductions in tourist arrivals* (*high confidence*). (IPCC 2014: 21; emphasis added)

It is in Chapter 12 of the IPCC AR 5 (IPCC 2014b) that we find a discussion focused on the 'migration and mobility dimensions of human security', citing a good deal of pertinent literature from social sciences. It is important to note that it is for the first time that the IPCC has decided to engage with human security issues and from a critical social science perspective this is a welcome move. While acknowledging that mobility and migrations have a complex geography (Banerjee and Samaddar 2006; Basu 2009) along with varied historical contexts and key drivers, it is pointed out that, 'As with other elements of human security, the dynamics of interaction of mobility with climate change are multi-faceted and direct causation is difficult to establish.'

One of the key findings of the IPCC 5 AR, which has a special resonance for this chapter is that places experiencing protracted conflict or recovering from conflict are much more susceptible and have fewer resources for dealing with weather extremes and climate variability.

Even though the IPCC presents itself as the international authoritative body on earth climate science, it has its critics and many criticisms relate to the politics of both knowledge production about climate change and related scenario building. Grundmann (2007: 416) points out that "contrarian' scientists and other critics think that the IPCC misrepresents the state of knowledge and exaggerates the size and urgency of the problem. While the skeptics accuse IPCC scientists of being environmentalists in disguise, others point to the processes of exclusion of specific social groups representing different knowledge claims'.

In contrast, James Lovelock (2010: 23) questions the deployment of the term 'consensus' on issues related to science and points out that 'it is a good and useful word but it belongs to the world of politics and the courtroom, where reaching a consensus is a way of solving human differences. Scientists are concerned with probabilities, never with certainties or consensual agreement.'

All said and done, Bangladesh, largely by virtue of its geographical/ geopolitical location, is widely perceived and reported as one of the most vulnerable countries to climate change-induced sea level rise (Brown 2009; Dodds et al. 2009; Faris 2009; Giddens 2009). Yet it is only by combining imaginative geographies of fear and hope that one of the leading and widely cited climate scientists like James Hansen (2010: 258), is able to convey the 'catastrophic' effects of climate change and sea level rise for 'developing' countries like Bangladesh:

> The consequences for a nation like Bangladesh, with 100 million people living within several meters of sea level, are too overwhelming, so I leave it to your imagination. No doubt you have seen images of the effects of tropical storms on Bangladesh with today's sea level and today's storms. You can imagine too the consequences for island nations that are near sea level. *We can only hope that those nations responsible for the changing atmosphere and climate will provide immigration rights and property for the people displaced by the resulting chaos.* (emphasis added)

It is difficult to deny that for coastal states such as Bangladesh, tropical cyclones, with major economic, social and environmental consequences, have always been a challenge. Countries that are most exposed for example, China, India, the Philippines, Japan, and Bangladesh have densely populated coastal areas, often comprising deltas and mega deltas (UNDP 2004). Every year, up to 119 million people are on average exposed to tropical cyclone hazard (UNDP 2004). From 1980 to 2000, out of a total of more than 250,000 deaths caused by tropical cyclones worldwide, nearly 60 per cent occurred in Bangladesh alone. Even this figure is less than the 300,000 killed in Bangladesh in 1970 by a single cyclone. Although 'The death toll has been reduced in the past decade due largely to improvements in warnings and preparedness, wider public awareness and a stronger sense of community responsibility' (ISDR 2004), as pointed out by the IPCC (see Nicholls et al. 2007: 337), Bangladesh remains one of the 'key hotspots of societal vulnerability in coastal zones' due to 'highly sensitive coastal systems where the scope for inland migration is limited.'

> Most of the land area of Bangladesh consists of the deltaic plains of the Ganges, Brahmaputra and Meghna rivers. Accelerated global sea level rise and higher extreme water levels may have acute effects on the human population of Bangladesh (and parts of West Bengal, India) because of the complex relationships between observed trends in SST over the Bay of Bengal and monsoon rains, subsidence and human activity that has converted natural coastal defenses (mangroves) to aquaculture. (Nicholls et al. 2007: 326)

Floods are among the most reported natural disasters in Africa, Asia and Europe, having affected nearly 140 million a year on average (Kundzewicz et al. 2007). In Bangladesh, three extreme floods have occurred in the last two decades, and in 1998 about 70 per cent of the country's area was inundated (Kundzewicz et al. 2007). In the case that global temperatures were to rise by 2°C, the area in Bangladesh to be flooded is likely to increase at least by 23–29 per cent (Kundzewicz et al. 2007: 187). So far, the efforts made by Bangladesh to put into place a sizeable infrastructure to prevent flooding have fallen short of desired results.

Against the backdrop of growing trends to securitize climate change-induced migrations in different parts of the world (Smith 2007), Barnett is quite right in pointing out that, 'The crux of the problem is that national security discourse and practice tends to appropriate all alternative security discourses no matter how antithetical' (Barnett 2003: 14). He also proposes that the IPCC scientists should downplay such climate change militarist discourses being 'cautious on the issue of violent conflict and refugees' and, instead, focus on climate justice issues. In his view, this approach 'might helpfully integrate science and policy and usefully elucidate the nature of the "danger" that the UNFCCC ultimately seeks to avoid' (Barnett 2003: 14).

We are quite in agreement with Barnett that the IPCC scientists should be extremely cautious while speculating over the geopolitical, strategic 'consequences' of climate. We would like to point out how a speech made by R. K. Pachauri, Chairman of the IPCC, and Director General, The Energy and Resources Institute (TERI), New Delhi, at the convocation of the Military college of Telecommunication Engineering, Mhow, on 26 June 2009, has been reported in the media under sensational headlines such as 'Global warming and how it encourages terrorism in

India' and 'Climate change your biggest enemy'. The press release of The Energy and Resources Institute (TERI 2009) reads as follows:

> Dr. Pachauri stressed on the global issue of climate change. 'Climate change poses new threats to India.' 'Melting snows in the north open up passages for terrorists, just as melting glaciers affect water supply in the subcontinent's northern part, sharpening possibility of conflict with our neighbours. Changing rainfall patterns affect rain fed agriculture, worsening poverty which can be exploited by others.' He added, 'Our defence forces might find themselves torn between humanitarian relief operations and guarding our borders against climate refugees, as rising sea-levels swamp low-lying areas, forcing millions of 'climate refugees' across India's border.

The science-geopolitics interface of climate change, as the above quotation appears to suggest, is rather complex, and as David Demeritt (2001, 329) succinctly points out,

> given the immensely contentious politics, it is tempting for politicians to argue that climate policy must be based upon scientific certainty. This absolves them of any responsibility to exercise discretion and leadership. This science-led politics is also attractive to some scientists since it enhances their power and prestige. However, this political reliance on the authority of science is deeply flawed: it provides neither a very democratic nor an especially effective basis for crafting a political response to climate change.

The Synthesis Report, the concluding document of the Fourth Assessment Report of the Intergovernmental Panel on Climate Change (quoted in IPCC 2010: 1) had stated:

> Climate change is expected to exacerbate current stresses on water resources from population growth and economic and land-use change, including urbanization. On a regional scale, mountain snow pack, glaciers and small icecaps play a crucial role in freshwater availability. Widespread mass losses from glaciers and reductions in snow cover over recent decades are projected to accelerate throughout the 21st century, reducing water availability, hydropower potential, and changing seasonality of flows in regions supplied by meltwater from major mountain ranges (e.g. Hindu-Kush, Himalaya, Andes), where more than one-sixth of the world population currently lives.

'Open borders to climate refugees': official claims and counter-claims in Bangladesh

In a speech delivered at the 64th Session of the United Nations General Assembly, on 26 September 2009, Sheikh Hasina, the Prime Minister of Bangladesh, talked about the implications of climate change for her country. It is worth noting that she not only makes a reference in this speech to millions of 'climate migrants' and 'climate refugees' but also emphasizes the need for an international legal regime for ensuring special social, cultural and economic rehabilitation of climate induced migrants. The tone and the tenor of her narrative of climate change, ably supported by natural science evidence, are visibly marked by the geopolitics of fear and cartographic anxieties:

> What is alarming is that a meter rise in sea level would inundate 18% of our landmass, directly impacting 11% of our people. *Scientific estimates indicate, of the billion people expected to be displaced worldwide by 2050 by climate change factors, one in every 45 people in the world, and one in every 7 people in Bangladesh, would be a victim.*
> Rapid, unplanned urbanization, occupational dislocations, food, water and land insecurity are some of the consequences of climate change. The affected communities would not only lose their homes, they would also stand to lose their identity, nationality, and their very existence, and in some cases, their countries (United Nations 2009; emphasis added)

It was in the same month (that is, September 2009) that the 'Bangladesh Climate Change Strategy and Action Plan 2009' was released. In her message to the Action Plan, Prime Minister Sheikh Hasina expressed the resolve of her government to 'free' the people of her country from the 'terror of climate' and to ensure that 'people are fully protected from its adverse impacts as promised in our manifesto' (Government of Bangladesh 2009: xi). However, the following excerpt from the report seems to suggest that the government of Bangladesh has not only accepted, rather uncritically, the category of 'environmental refugees' but has also embraced the fear-driven geopolitical assumption that 'more than 20 million' displaced Bangladeshis will be migrating to other parts of the world:

> It has been estimated that there is the impending threat of displacement of more than 20 million people in the event of sea-level change

> and resulting increase in salinity coupled with impact of increase in cyclones and storm surges, in the near future. The settlement of these environmental refugees will pose a serious problem for the densely populated Bangladesh and migration must be considered as a valid option for the country. Preparations in the meantime will be made to convert this population into trained and useful citizens for any country. (Government of Bangladesh 2009: 17; emphasis given)

The quotation above raises a number of intricate and intriguing questions (Doyle and Chaturvedi 2011). Who are environmental refugees? Who among the imagined millions of displaced Bangladeshis, 'in the event of sea-level rise', for example, would qualify to be 'environmental refugees' and why? Is there no difference whatsoever between 'climate refugees' and 'environmental refugees'? What are the grounds for assuming that those displaced due to either sudden onset of natural disasters or gradually unfolding climate change or even abrupt climate change would necessarily choose to cross the borders in search of safer and greener pastures? Elizabeth G. Ferris (2008: 83), has argued that 'It is also likely that most of those displaced by these types of events will remain within their country's borders.' Who would decide what kinds of preparations are needed to turn the displaced millions 'into trained and useful citizens for any country'? Why can't these potential 'Others' be trained into useful work force as citizens of Bangladesh; their homeland?

'UK should open borders to climate refugees': this is how the Guardian newspaper (see Grant et al. 2009) reported the first ever alarmist statement by any senior politician of Bangladesh (the finance minister, Mr. Abdul Maal Abdul Muhith), just before the Copenhagen climate summit (COP 15). He emphatically pointed out the moral responsibility of Britain, the USA and other countries of the global North to accept climate refugees from Bangladesh. Mr. Abdul Muhith is reported to have told the Guardian, 'Twenty million people could be displaced [in Bangladesh] by the middle of the century ... We are asking all our development partners to honor the natural right of persons to migrate. We can't accommodate all these people, this is already the densest [populated] country in the world' (ibid.). Curiously enough, he also underlined the need for the UN to redefine international law in such a manner that climate refugees are provided with the protection at par with people fleeing political repression. Echoing the assurance provided in the Bangladesh Climate Change Strategy and Action Plan (2009), pointed out by us above, Abdul Muhith expressed the hope that if properly 'managed' inter-state migration could be positive for both Bangladesh and the west:

> We can help in the sense of giving the migrants some training, making them fit for existence in some other country ... Managed migration is always better we can then send people who can attune to life more easily ... Total aid in Bangladesh today is less than 2% of GDP. It is almost the same in China and in India. So we, the most populated, least developed country, get peanuts. This inequity is terribly intolerable. (Grant et al. 2009)

Whereas the concept of 'managed migration' mentioned above remains alarmingly vague, the response of Rajendra Pachauri, Chairman IPCC, to the statement made by the finance minister of Bangladesh, and quoted by *The Guardian*, also raises a number of intricate issues:

> This is clearly a warning signal from Bangladesh and similar countries to the developed countries. And I think it has to be taken very seriously. If you accept that those countries that have really not been responsible for causing the problem, and have a legitimate basis for help from the developed countries, then one form of help would certainly be facilitation of immigration from these countries to the developed world ... If you had 30 or 40 million migrating to other parts of the world, that's a sizable problem for which we have to prepare. And if it requires changes to immigration laws and facilitating people settling down and working in the developed countries, then I suppose this will require legislative action in the developed world.

It is to state the obvious, perhaps, that if (and this is a big if) 30 or 40 million climate migrants were to cross international borders, that will be a 'sizeable problem (but for whom?) for which we (does this "we" imply the host country like the UK or USA, the developed West, or the so-called, geographically undifferentiated international community) have to prepare', and that would need, among other things, huge sums of money. And in his response, Douglas Alexander, the international development secretary of the UK was quick to point out: 'As the largest international donor to Bangladesh, Britain has been urging the international community to provide extra money for climate change adaptation' (Grant et al. 2009). The Guardian also quoted Jean-Francois

Durieux, Deputy Director, Division of Operational Services at the United Nations High Commissioner for Refugees (UNHCR), as having said,

> The risk of mass migration needs to be managed. It's absolutely legitimate for Bangladesh and the Maldives to make a lot of noise about the very real risk of climate migration they hope it will make us come to their rescue. But reopening the 1951 convention would certainly result in a tightening of its protections ... The climate in Europe, North America and Australia is not conducive to a relaxed debate about increasing migration. There is a worry doors will shut if we start that discussion. (Grant et al. 2009)

The Guardian story provides a useful insight into the fact that climate change rhetoric often deploys a calculated politico-legal ambiguity, depending upon the interests of the actors and agencies involved, in order to hide the underlying anxieties and fears. It is to the legal dimensions of the geopolitics of fear that we turn next. However, before we do, it will be instructive to take note of the fact that the imaginative geographies of climate change-induced, trans-border migrations are also creating considerable cartographic anxieties in the immediate neighborhood of Bangladesh, especially India.

A major contribution to the special issue on climate change of *Himal South Asian* (Chowdhury 2009), a leading and widely read scholarly magazine published from Kathmandu, Nepal, refers to various facets of fast ascending geopolitics of fear about 'climate refugees' in South Asia. The narrative begins by posing a couple of questions. 'Could India afford to refuse sanctuaries to Hindu (climate change) refugees from Bangladesh, which would certainly be a demand of many Hindu Indians? And if the state were to do so, how would Indian Muslims react?' (ibid.). The argument then becomes that were India to discriminate in favor of Hindu refugees, millions of Muslims, including Bengali Muslims in India, would seriously doubt the secular credentials and claims of the Indian State. Moreover, such a policy choice would further strengthen the hands of various regional and global radical, right wing organizations and their propaganda networks. Given India's ongoing battle against cross border militancy, 'If it were to begin to refuse entry to climate change refugees, the country would suddenly have to face a far larger extremist problem from both and with a far larger hostile population within, to boot.'

And were India to decide in favor of letting even some refugees in on certain grounds, including humanitarian, the prevailing social-political sentiment in the Indian northeast, (bordering on resentment) against Bangladeshi/Muslim refugees and migrants would become more pronounced. This may even result in more frequent attacks on refugees, provide a further fillip to ongoing insurgencies, and lead to further proliferation of the perception that the Indian state is unable to secure the communities of the northeast against the new influx of refugees. 'The echo effect across the border would likely escalate the crisis.' Chowdhury (2009) concludes on the note that come what may, India simply does not have the 'capacity' to cope with the 'flood' of millions of 'climate change refugees'. Questions such as the following remain unanswered in his view: 'Living next door to Bangladesh, can India handle the problems, apart from challenges in the context of climate change? It may not fail but could it prevent itself from being overwhelmed? And if India were to be overwhelmed, could the region survive in any conventional sense?' (ibid.)

Issues raised by Chowdhury acquire significance and salience in the light of statements made by India's new BJP Prime Minister, Narendra Modi, during his election campaign in India's Northeastern state of Assam. He was reported by *The Times of India* (24 February 2014) to have said:

> As soon as we come to power at the Centre, detention camps housing Hindu migrants from Bangladesh will be done away with ... We have a responsibility toward Hindus who are harassed and suffer in other countries. Where will they go? India is the only place for them. Our government cannot continue to harass them. We will have to accommodate them here.

Geopolitical fears and legal hopes: rhetoric and realities

B.S. Chimni (2000: 1) has argued that, 'the definition of a "refugee" in international law is of critical importance for it can mean the difference between life and death for an individual seeking asylum.' According to the 1951 Protocol and Convention Related to the Status of Refugees (Hathaway 2005), subscribed to by more than 100 states, a refugee is one who,

> as a result of events occurring before 1 January 1951 and owing to well-founded fear of being persecuted for reasons of race, religion, nationality, membership of a particular social group or political opinion, is outside the country of his nationality and is unable or,

> owing to such fear, is unwilling to avail himself of the protection of that country; or who, not having a nationality and being outside the country of his former habitual residence as a result of such events, is unable or, owing to such fear, is unwilling to return to it. (quoted in Chimni 2000: 2)

It is to state the obvious, perhaps, that unless it is accepted by all concerned that 'nature' or 'environment' or 'climate' can be the persecutor, the term refugee, as subscribed to by the 1951 Convention, is not going to work for those supposedly or actually displaced by natural disasters or climate change (Renaud et al. 2007: 14). It is worth noting that even though the temporal and geographical restrictions imposed by the 1951 Convention on the definition of 'refugee', against the Cold War politicking, were removed by the Protocol of 31 January 1967 relating to the Status of Refugees, what remained by and large untouched was fear as the key defining principle along with its Eurocentric bias. According to Chimni (2000: 7), 'this meant that most third world refugees continued to remain de facto excluded, as their flight is frequently prompted by natural disaster, war or political and economic turmoil rather than by "persecution", at least as that term is understood in the Western context.' It is equally noteworthy that regional instruments such as the 1969 Organization of African Unity (OAU) Convention and the 1984 Cartagena Declaration, besides incorporating region-specific attributes into their expanded definitions, went on to emphasize that the categorical understanding of a refugee should move away from a geopolitically dictated principle of 'well-founded fear' of 'persecution' to address the plight of those fleeing civil unrest, war and violence, irrespective of whether or not they can prove a well-founded fear of persecution.

The point we are trying to drive home is that despite such attempts to broaden and deepen the understanding of the term 'refugee', a categorical approach that is deeply embedded in the notion of 'fear' appears rather overwhelming in this entire enterprise. Be it the 'subjective fear approach' (based on a refugee's own assessment of the risk he or she faces) or the 'objective fear test' (ensuring that refugee's subjecting risk assessment must not be contradicted by the 'objective' circumstances of the case), it is the notion of fear that constitutes the fulcrum around which various competing definitions of refugee seem to revolve. What seems to be further magnifying the fear factor is the sheer numbers game, which, depending upon the actor and/or agency citing certain figures, seems in turn to be marked by calculated ambiguity. According to Myers (2002; 2005), as compared to 25 million people who migrated in 1995, there would be around double that figure by 2010, and by the end of the 21st century nearly 200 million would migrate due to climate change. What the tyranny of numbers ably hides is the question of who will choose to migrate, when, where and how far?

Our intention here is not to dismiss outright the notion of fear as of no moral or practical value in answering the question as to who is a refugee. What we do intend is to raise the following question: in various imagined or actual encounters between those seeking the refugee status and those who had the authority to grant that status, whose 'fear' is likely to dictate the process and decide the outcome?

Existing legal structures, such as the Refugee Convention and the Framework for Internally Displaced Persons (IDPs), were built for disparate reasons and, at best, have been limited in their application; at worst, these structures are fundamentally inept within the climate context. An alternative is a regionally defined regime operating under the rubric of the UN Climate Change Framework. Williams argues that although

> the Climate Change Convention and the Kyoto Protocol currently call for regional cooperation regarding adaptation activities, it is argued there should be an explicit recognition of so-called climate change refugees in the post-Kyoto agreement that allows for, and facilitates, the development of regional programmes to address the problem. Such a strategy would remedy the current protection gap that exists within the international legal system, while allowing states to respond and engage with climate change displacement in the most regionally appropriate manner. (Williams 2008)

(Moberg 2009: 1107), on the other hand, maintains that, although governments could utilize current international and domestic definitions of refugee to protect environmentally displaced persons, 'it is unlikely that any government will do so.' Even if they did 'extend these existing refugee and asylum laws to include environmentally displaced persons', the protection would be inadequate. It would also 'consume judicial resources needed for persons currently receiving protection under refugee and asylum laws'. Instead, Moberg argues, new domestic and international laws should be made in order to put environmentally displaced persons under a more 'protective, cost-sharing approach' (ibid.).

Moberg also suggests that environmentally displaced persons (EDPs) should be granted protection under their own Environmentally Based Immigration Visa (EBIV) Program: 'Similar to the current refugee program, countries should share the burden of accepting and supporting

EDPs, with that burden resting more heavily on wealthier nations' (Moberg 2009: 1135).

Conclusion

As argued earlier in this work, we have no intention either to deny the science of global warming (although we do feel that expressions such as 'certain' and 'consensus' go against the very nature and purpose of science and scientific pursuits), or to dismiss wide-ranging implications of climate change for the state, society and, in this specific chapter's case, Bangladesh. What we have questioned with all emphasis at our command, however, is the geopolitics of climate fear and the underlying alarmist imaginative geographies of 'climate migrants' and 'climate refugees'. These imaginative geographies of doom, disaster, and development, framed and flagged at various sites in the service of diverse agendas, are increasingly shaping the world-view *of* and *on* Bangladesh. We have tried to show how fear is aroused and mobilized in the service of various agendas of political, economic and cultural controls. We have also shown how, in the process of the national and local priorities becoming skewed, the discourses of mitigation and related structural approaches (with concomitant aid seeking strategies) overshadow adaptive strategies to climate change, especially migration. The failure to arrest and reverse such trends might result in what Foster and Clark (2009: 260) describe as the 'Fortress World' with its 'protected enclaves'; '... a planetary apartheid system, gated and maintained by force, in which the gap between the global rich and global poor constantly widens and the differential access to environmental resources and amenities increases sharply.'

This chapter has further argued that the meaning, nature and scope of 'climate change' discourses need to be broadened and deepened much beyond the science, ethics and politics of 'global warming' and its various manifestations such as the melting of polar icecaps, glaciers and rising sea levels. There is a need to acknowledge that global warming and its several facets labeled 'climate change' are, no doubt, a compelling, but not the only, issue on the agenda of environmental security in the Indian Ocean Region (Doyle 2008). Climate change is neither a moment of rupture nor departure (although it is often made out to be so) in the long-standing history of ecological destruction and deeply entrenched ecological irrationalities in modern-capitalist societies. In the absence of such an acknowledgment, the climate change discourse becomes both limited and limiting.

Our analysis of complex ground realities in Bangladesh also underlines the fact that 'there is ample conceptual muddle regarding CIM'

(White 2011: 29) and furthermore 'the invocation of present and future refugee crises stemming from climate change can be politically loaded, and the potential for political misuse is profound' (ibid.). Acknowledging the presence of global North-South dichotomies in human rights, equity and social justice dimensions of climate change, our engagement with Bangladeshi discourses on these issues shows that the issue of CIM in particular cannot (and should not) be divorced from the complex geopolitical and social-ecological entanglements and contestations. The 'innocence' of the power-political elite in the global South (as in the case of Bangladesh) cannot be taken for granted, nor the addition of 'climate walls' to existing assemblage of walls/barriers be overlooked.

Our discussion of competing fear-inducing imaginative geographies of climate change at various sites, and the manner in which Bangladesh is being implicated in them, reinforces Gregory's insightful comment that, 'imaginative geographies are spaces of constructed (in)visibility and it is this partiality that implicates them in the play of power' (Gregory 2009: 371). Imaginative geographies of 'coming climate catastrophe' (Hansen 2010) make the long history of ecological degradation and ecological irrationalities, perpetuated by the economic growth oriented models of development (and further legitimized by the powers that be in the name of 'national interest' and 'national security') almost invisible. On the other hand, 'diverse environmental problems have essentially been laid at the door of climate change ... New problems have been grafted onto old ones and given a single cause; an example of a "garbage can anarchy", where once isolated phenomena become systematically interrelated' (Connell 2003: 98).

Having noted that, our analysis of the dynamics and dilemmas of climate change in the case of Bangladesh reveals that, depending upon their power-political moorings and power-political agendas, geopolitics of climate change will continue to oscillate between imaginative geographies of fear and hope. Equally important, however, is the geopolitics of humiliation to which the displaced communities in Bangladesh are likely to be increasingly subjected, both discursively and on the ground. The geopolitics of fear, articulated through highly alarmist imaginative geographies of climate change-induced displacements, will continue to discursively displace the equity-based notion of 'ecological debt' by the consequence-based, neoliberal notion of 'carbon footprint'. We fully agree with the wise contention that the 'fear agenda' should be questioned and challenged so that the media and the governments do not incite unhelpful and inaccurate slogans on immigrants and the comprehension of the term 'migrant' or 'refugee' should be expanded to that which meets his/her human security (Gupta 2009: 77). After all, there are millions, and a vast majority of them in the Indian Ocean Region, who live in fear of their 'human security' being seriously compromised but without giving up the hope of a better, just and humane 'world order', with or without a changing climate. We have argued that in relation to climate change-induced displacements and migrations, in the case of Bangladesh and its immediate neighborhood, fear rather than hope seems to be growing, fuelled by Northern-centric cartographic anxieties. But, as we began with Moïsi's thoughts on the geopolitics of fear, so too, we wish to conclude by revisiting the mirror-opposite of this emotion: hope.

6
Climate Security and Militarization: Geo-Economics and Geo-Securities of Climate Change

Introduction

Although there have been conflicts over resources since the earliest human societies, interest in both renewable and non-renewable resources within environmental security frameworks has dramatically increased since the end of the Cold War (Doyle 2008). Security is usually understood in state-centric terms, 'concerned with intentional physical (mainly military) threats to the integrity and independence of the nation-state' (Scrivener 2002: 184).

This trend was reinforced and supported by military establishments who sought means by which they could continue to justify Cold War levels of military expenditure during an apparent time of peace and prosperity for the West. As a result, both the US and Russia formed high level units of environmental security within established security institutional infrastructure, such as the Pentagon's Centre for Environmental Security. Paradoxically, some peace advocates also championed the concept. In a book appropriately entitled *Green Security or Militarized Environment*, Jyrki Kakonen writes:

> Peace researchers have argued for environmental security in order to show that ... national defense resources could be used for civilian purposes in the field of environmental problems. The aim is to convert military resources ... to do the environmental protection in order to transform the military into a paramilitary and further into a non-military organization. This is an option after the Cold War, but there is a danger that the militarist approach to deal with environmental issues leads to the militarization of the society. (Kakonen 1994: 4)

As with the now dominant concept of *terrorism*, *environment* and *climate change* are extremely powerful multilateral concepts which cross nation-state borders relatively easily. In this vein, a new enemy is imagined, one which was not a human terrorist, but *nature* itself. This interest in a *combative environment* is not new. In western terms it has been aptly recorded in the 18th Century works of Thomas Hobbes. In his most famous work – *Leviathan* – Hobbes depicts nature as in a state of perpetual war with itself (Hobbes 1651). This conservative view of nature was used by Hobbes to justify his call to create an all-powerful authoritarian 'machine' which would be the only means to avert global environmental catastrophe. Of course, in this understanding of nature, humanity is also in a perpetual 'state of war' (ibid.). This western understanding of the 'state of nature' is not just restricted to neo-Hobbesians, but has substantial populist credence, as most western imaginings comprehend peace to be the aberration whilst war is construed as the 'natural state'.

This interest in Gaia as the common enemy (as well as a pathway to common salvation) was further heightened in the West due to increasing, but rather late, understandings that the minority world (the more affluent world) had to share its basic survival systems with the majority world (the less affluent world). This concept of a shared spaceship Earth had been vociferously pushed by the western environmentalists since the late 1960s, but due to characteristic conservatism of traditional security studies, this green rhetoric was only picked up in the academic literature and the governmental grey documentation in the late 1980s and early 1990s.

'Environmental security' emerged forcefully in the Brundtland Report in 1987, and increased at the first Earth Summit in Rio de Janeiro in 1992. The nexus between environment, development and security was never stronger than at the 'Earth Summit Plus Ten' in Johannesburg in 2002 (Doyle 1998). The notion of environmental (and climate) security, however, is hotly contested. Its most common variation is concerned with the impact of environmental stress on societies, which may lead to situations of war within and between societies. In this manner, environmental security agendas are about seeking issues which, if not addressed, may provide the basis for increasing human conflicts. In this sense, environmental and/or climate security is understood in somewhat negative terms.

In the Westphalian mindset, the advantage of reinstituting nature as the redeclared enemy was that it constituted a 'common' security issue for all humankind, or in the words of Brundtland – 'Our Common Future'. Of course, the symbol 'environment' is not common at all;

rather the issues which gather under its umbrella are culturally diverse. Concepts of environment (and the later variant, climate change) are far from apolitical; rather, they are the exact opposite (Doyle 2002). They are intensely politicized categories utilized to redraw boundaries of collective identity, behavior, political activity, security and, most importantly, power and resource distribution.

Securitizing and militarizing climate change across the ideological divide

Despite the prevailing dominance of 'realist' schools of thought in their interpretation, climate change – now cast within security frames – has proved very successful in its appeal all along the ideological sliding scale. Climate change's omniscience is well reflected in security debates, predominantly from different positions on the ideological Right (but with some belated input from the Left). These positions match neatly with the ideological tussle between the traditional liberal and neo-liberal positions articulated in Chapter 4. The strong connections between these ideologies using both economics and security as their nexus points makes great sense here, as the majority of this chapter explains how neo-liberalism, in the end, is a key driver in promoting climate change from not just a *threat multiplier* to a far more powerful *force multiplier*. Indeed, under the umbrella of anthropogenic climate change, the connections between science and technology, markets and security have never been stronger.

In a bid to understand these diverse security frames and, indeed, to comprehend how human-induced climate change has been securitized, we will now briefly visit three major schools of thought within geo-politics and international relations. We then examine the manner in which they have interpreted climate change problems (and environmental issues in general), and the ways in which they are imagined to be addressed. As always, these theoretical categories are far from perfect, and often both descriptions and prescriptive policies to avert climate change blur across these conceptual boundaries.

First and foremost, the 'traditionalists' tend to come from the 'realist' school. Not only do they see climate issues through the prism of 'national interests' within the context of an anarchical world system; they see climate as just another issue pertaining to the struggle for power amongst nation-states. In addition, climate, as a form of environmental security, is usually seen as a 'threat-multiplier,' rather than a base or fundamental threat. In this vein, climate can exacerbate tensions but, as an 'alternative' form of security (and, therefore, not as a 'fundamental' one of race, religion, ethnicity, finances etc.), it acts as an 'accelerator' or 'catalyst' for existing tensions between nation-states (for examples of this dominant realist – and usually militarist – approach, see Myers 1993; Salehyan 2005; Reuveny 2007; Chin 2008). Interestingly, not only does it imagine an anarchical world system; it views natural processes themselves as anarchical. In this world view, humanity has not only declared a war against itself, but is also locked into mortal combat with the earth itself – Nature as enemy. In a recent book by Gwynne Dyer, this realist version of environmental security is portrayed nicely in *Climate Wars* (2008), echoing neo-Malthusian and neo-Hobbesian sentiments.

In this view of climate security, nation-states are seen as having to protect their borders from climate refugees (see Chapter 3) driven from the global (and particularly global South periphery); protecting their 'natural comparative advantage' (in Ricardo's terms) of coal, uranium and other markets. Ever since the concept of 'global climate change' arose to prominence in the 1980s, a series of metaphors have been deployed at the service of imaginative geographies of chaotic and catastrophic consequences of climate change, including 'mass devastation', 'violent weather', 'ruined' national economies, 'terror', 'danger', 'extinction' and 'collapse'. A number of 'security' experts and analysts are convinced that the United States will be the 'first responder' to numerous 'national security' threats generated by climate change (see Podesta and Ogden 2007). In April 2007, the CNA Corporation (2007), a think tank funded by the US Navy, released a report on climate change and national security by a panel of retired US generals and admirals that concluded: 'Climate Change can act as a threat multiplier for instability in some of the most volatile regions of the world, and it presents significant national security challenges for the United States.'

One of the key pressure points in the realist and largely militarist re-constructions of environmental issues (using climate as a political metaphor) relates to the 'blood-dimmed tides', the washing up of climate refugees – and even climate terrorists – upon the shore of the affluent world, due, in part, to rising sea-levels. If anything, this imagined ecological Armageddon sees the global North re-engaging with the global South, not through choice, but through necessity (ibid.: 1).

More liberal notions of international relations re-emerged after the so-called 'victory of capitalism' and the breakup of the communist-inspired USSR in the late 1980s, and world orders which had existed since World War II were called into question. During this time of uncertainty, there emerged a global, almost post-modern, policy-shaping concept embracing a shared plurality of interests which crossed nation-state borders, commonly referred to as *multilateralism*. The multi-lateralist decade of the 1990s, which ended as the current phase of US unilateralist emerged forcefully in the new millennium, was an era when new boundaries and borders were drawn in the sand, as alternative, more liberal concepts of identity and collectivity were imagined. One such idea which evolved at this time was that of *environmental security* (Doyle and Risely 2008) and later, it most powerful and neo-liberal off-spring, *climate security*.

In a positive light, a more liberal concept of environmental and climate security which is more inclusive of the interests of the majority of people in the global South is one that moves away 'from viewing environmental stress as an additional threat within the (traditional) conflictual, statist framework, to placing environmental change at the centre of cooperative models of global security' (Dabelko and Dabelko 1995: 4). In these terms, as discussed in Chapter 4, the Kyoto protocol is just one powerful example of how the affluent North seeks to re-engage with the less affluent world, through more co-operative, bi- and multi-lateral negotiations. Until very recently, of course, countries such as the United States and Australia positioned themselves well outside these co-operative ventures, maintaining a nation-state centric regional and global stance.

A good example of a liberal version of the securitization of climate lies in the political realm of non-state actors or organizations. For example, E3G is a new organization that aims to convert environmental goals into 'accessible choices'. Its CEO is John Ashton, who co-founded it with Tom Burke, an academic and former special advisor to three UK Secretaries of State for Environment. Burke and Ashton argue that:

> Climate change is not just another environmental issue to be dealt with when time and resources permit. A stable climate, like national security, is a public good without which economic prosperity and personal fulfillment are impossible. It is a prime duty of a government to secure such goods for their citizens. The current level of investment of political will and financial resources addresses climate change as an environmental rather than as a national security issue. Without a fundamental change in this mind-set governments will remain unable to discharge their duty to their citizens. (Burke and Ashton 2004: 6)

Liberals, in security terms, often focus on mitigation, which includes programs and policies that concentrate on reducing ecological footprints (today). Through promoting and working within co-operative international conventions and protocols, like Kyoto, there are obvious efforts to mitigate against climate change.

Finally, there are subservient, but more critical traditions within the rhetoric of environmental security. We have space to investigate only two categories of argumentation here. Advocates of the first type of critical position do not concentrate, as their liberal counterparts do, on ecological footprints, but rather upon ecological debts incurred over centuries of exploitation of the North over the South. This position is still crafted within discourses of security, focusing on the *causes* of environmental *in* securities (Barnett 2003). For example, in the Asian region afflicted by the Tsunami in 2004, both realist and liberal responses were used as justifications to deploy military personnel to the worst affected areas, or to construct co-operative 'early warning systems' across the oceanic region. On the other hand, a critical response reviews and responds to the causative factors leading to environmental instabilities and insecurities, seeking, amongst other things, reparations for climate-vulnerable communities.

The approach of transnational green NGO Friends of the Earth International (FoEI) is of interest here (this case study will be developed in full in Chapter 7). In a report made by the organization, Davissen and Long make FoEI's critical position clear:

> The global North, as the major greenhouse polluters, bears a significant responsibility for this disruption. Accordingly, we believe that the North must make reparations. (Davissen and Long 2003: 8)

Climate debt is the special case of environmental justice – where industrialized countries have over-exploited their 'environmental space' in the past, having to borrow from developing countries in order to accumulate wealth, and accruing ecological debts as a result of this historic over-consumption (ibid.). Friends of the Earth (see Chapter 7) maintain that the 'climate debt' owed by the global North to the global South is accruing due to the 'unsustainable extraction and consumption of fossil fuels', and therefore, climate debt is a key component of a more equitable form of environmental security. Consequently, FoEI argues that the responsibility for climate change and accommodating past earthly indiscretions rests firmly on the minority world.

Whilst some commentators and organizations like FoEI seek to introduce critical arguments *within* the dominant climate change frame, another critical approach is to question the validity of the environmental security framework itself. As mentioned, this latter critical approach largely informs our own stance on the matter (Chaturvedi and Doyle 2010). From this position, due to the fact that climate change discourses have already been populated – and ultimately dominated – by more realist and neo-liberal (and even classical liberal) discourses, climate-displaced persons do not gain through being included in this narrative. Despite FoEI's best emancipatory intentions, they will, at best, achieve little within this frame and, at worst, legitimate a climate discourse which further securitizes the global South within a social and environmental agenda initiated, shaped and controlled by the global North (see Chapter 7).

Within this latter type of critical approach, writers such as Simon Dalby also argue that the very act of securitization, with its implicit agenda set by military-industrial complexes, ultimately disenfranchises the majority, stripping environmental 'speech' from its more emancipatory projects (Dalby 2002).

Jon Barnett attempts to provide some way out of what can become a critical 'dead-end'. Like other critical geopolitical thinkers, he does acknowledge that, 'The crux of the problem is that national security discourse and practice tends to appropriate all alternative security discourses no matter how antithetical' (Barnett 2003: 14). But he also attempts to provide real policy choices: he proposes that the IPCC scientists should downplay such climate change militarist discourses – being 'cautious on the issue of violent conflict' – and, instead, focus on climate justice issues. This approach, he argues 'might helpfully integrate science and policy and usefully elucidate the nature of the "danger" that the UNFCCC ultimately seeks to avoid' (Barnett 2003: 14).

What has emerged from our research is the clear knowledge that science and technology, economics and security have merged. For the remainder of this chapter we investigate these connections, and for the purposes of this discussion, we focus on the Indo-Pacific region (IPR). We choose this region (or non-region) as a rapidly emerging 'post-political' space, which accounts for approximately 60 per cent of the Earth's people, and is also the current powerhouse of the global economy and equally essential to the Earth's security. Although the rhetoric of realism underwrites much of the literature and discussion of environmental security, newer, more neo-liberal forms of securitization and militarization are emerging forcefully across the globe in the name of *climate security*.

Climate change as comprehensive security in the continuum: geostrategy and geoeconomics in the Indo-Pacific

As global faultlines multiply and variegate in the 21st Century, new regional networks and constellations are proliferating, each one built on alternate constructions of geoeconomic and geopolitical (in)securities. The 'rise' (and return) of the 'Indo-Pacific' region has brought with it new and contested forms of map-making, spatially redefining the relationships between domestic and regional governance, alongside transnational markets. Countries within the region usually 'mark out' the region using more traditional, though differentiating, territorial borders to suit their geoeconomic and geostrategic needs. Borders and boundaries, therefore, are shuffled appropriately. This chapter, however, focuses on a new form of 'region-building' (or deterritorialization) which is, in fact, the creation of the Indo-Pacific as a *continuum, super-region*, or *non-region*. This version of the IPR has emerged as one dominant but competing view, largely promulgated within United States Defense circles. It becomes particularly powerful in both geostrategic and geoeconomic terms when coupled with the dynamics and markets of global climate change which provide the perfect narrative space for this neoliberal, 'de/reterritorialization'. In this 'new' space, the US military has embraced the climate bandwagon (in fact, it has become one of climate change's biggest supporters in financial terms). Climate re-endorses the military's strategic entry into all places and spaces, as climate change is constructed as omniscient, and an enemy to all – usually constructed as a *conflict-multiplier* (in a realist sense). But better than this, although sometimes contextualized within these green moral arguments, it makes both economic and security sense, as climate is also constructed by the military through the lens of energy security and, in doing so, morphing into a *force-multiplier.*

Obviously, right at the start of this argument, we have to make a case for a broader understanding of security than what is usually meant when using narrower, more traditional security theoretical frameworks in politics and international relations (IR). Security in IR is now widely considered more inclusive than 'guns and tanks', and even within traditional 'statecraft' literature, links between economic and military security are now well-established (see Mastanduno 1998).

In more traditional political science, political geography and international relations circles, there has existed a rarely questioned, widely understood assumption that geoeconomics and geosecurities were clearly separated in both a conceptual manner, and in terms of 'real' deployments of power (Doyle and Alfonsi 2014). In more recent times, however, it is more acceptable to argue that economics and politics (particularly in security discourses) are integrally (and quite obviously) related. Indeed, these connections in terms of realpolitik have been with us for some time now. As Michael Mastanduno (1998) has argued: 'the strategic urgency and uncertainty perceived by the U.S. officials in the early Cold War setting led them to integrate economic and security policies. U.S. officials sought to create an international order, and economic instruments and relationships were a vital part of that undertaking.' This argument pertaining to the increasing inseparability of geopolitics and geo-economics – particularly during times of rapidly changing cartographies of power – is central to this chapter.

Within non-traditional security circles, we draw on just two of these broader definitions of security here. First, we utilize the work of Luciani (1989: 151), who writes that, 'National security may be defined as the ability to withstand aggression from abroad' but discusses that within the context of economic, and not just territorial aggression. This is particularly pertinent when discussing the power of transnational corporations (TNCs) (and nations such as the US, Japan and China with large stakes in TNCs) and their relationships with nation-states. Secondly, the thoughts of Ullman (1983: 133) are also most apposite within this context:

> A threat to national security is an action or sequence of events that (1) threatens drastically and over a relatively brief span of time to degrade the quality of life for the inhabitants of a state, or (2) threatens significantly to narrow the range of policy choices available to the government of a state or to private, nongovernmental entities (persons, groups, corporations) within the state.

This definition of security is a vital premise for later arguments on economic, human and environmental insecurities in nations within the region, and across the IPR as a whole.

Although the Cold War, as Mastanduno argued, saw connections between economy and geopolitics (in security terms) emerge forcefully, the entrenchment of neo-liberalism in the decade of US-led multilateralism which followed the demise of the Soviet Union further entrenched both the discourses and the realpolitik and real markets. In economic terms, particularly from the 1980s, global financial institutions implemented major macro-economic restructuring, and embraced the so-called 'open economy'. These economic reforms included: the floating of local currencies against the US dollar; the reduction of international tariffs to facilitate international trade; the deregulation of national financial systems; the relaxation of controls on foreign investments; the reduction of government spending; and the privatization of many state-owned enterprises. Labour reforms such as enterprise bargaining and other measures to curb the power of labor were subsequently also introduced in many countries both across the globe and within the IPR (Doyle and Alfonsi 2014).

The free trade, open market reforms implemented in the 80s and 90s are now being implemented around the world – even in the airways and seaways of transnational space – the mantra of governments everywhere. As Stedman Jones (2012) has argued, there was nothing inevitable about this. The international consensus governing the rules of international trade and finance which emerged from the Bretton Woods regime after the Second World War was seemingly in disarray by 1970. Persistent stagflation, unemployment and increasing labor unrest undermined confidence in the Keynesian theory that the state could and should intervene in markets in order to guarantee social stability and welfare, securing employment and economic growth.

While opponents of the Bretton Woods consensus, such as the Chicago School of economists, argued that state intervention was indeed at the core of economic problems rather than being the solution, there were also those on the left who felt radical policy change was necessary (ibid.). In Australia (an aspiring middle power within the IPR), these reforms were carried out with the support or compliance of the trade union membership who, some commentators would argue, have lived to regret their role in this ideological and policy shift (Cahill 2008).

Economics as transnational security

Neo-liberal reforms to allow free reign to the markets, however, were not merely economic dogma. To one of neo-liberalism's chief architects, Frederic Hayek, neo-liberalism was as much a social and moral philosophy as an economic doctrine. State economic planning, for example, is not only wrong in economic terms – it is also inherently the road to totalitarian control over societies and individuals. Human happiness, according to Hayek, is best achieved if people are left to pursue their own interests and goals. State intervention on economic activities could only eventuate in coercion and oppression, and hence is inherently unjust (Hayek 1944).

The only role for the state, according to this normative position, is to provide strong protection for property rights and to facilitate the conditions by which individual property rights and markets can flourish. As promoted by politicians like Thatcher and Reagan, therefore, neo-liberalism was presented as a great utopian project that secured the 'noble' aim of individual liberty; the maximization of wealth and democracy, and ultimately, a stable international order.

After the collapse of the Soviet Union, neo-liberal ideology has emerged as the triumphant hegemonic discourse – a kitbag of doctrines, values, and policy prescriptions – applied globally by extremely powerful international institutions, regimes (such as the International Monetary Fund and the World Bank), as well as by private corporations. The benefits of this project, however, are contestable. To its detractors, the primacy of markets is merely the latest manifestation of class-based capitalist oppression. Neo-liberalism pays little more than a lip-service in its promotion of human liberty; rather, it is often a seductive disguise hiding more modern forms of neo-colonial expansion and exploitation (Harvey 2005; Klein 1999; 2007). As Harvey notes, when there is a clash between the principles and the specific interests of global elites, it is the latter which are protected (Harvey 2005: 19) – a point surely made very clear after the events of the global financial crisis (GFC) (Gamble 2009).

Critics of neo-liberalism, alternatively from the neo-Marxist left or the neo-mercantilist right (O'Brien and Williams 2010), point to the increasing inequality it engenders between the rich and the poor; its undermining effects on national sovereignty and democracy; how it has come to pervert public goods which should never be marketized (Sandel 2012); and, finally, how marketized values and politics have entrenched a new super plutocracy, one which is essentially transnationalized, largely impervious to the limitations of national borders (Freeland 2012). Stedman Jones (2012) charts how – despite the fault-lines and ruptures underlying the GFC – it has become a blind faith, a mantra that enmeshes individuals, societies, and governments, securing some powerful minority interests whilst making the lives of the global majority more vulnerable.

Mapping the 'Indo-Pacific': the creation of the US-led continuum

Persisting uncertainties over the spaces and spatialities of the 'modern era' and the corresponding contestation over the cartographies of globalization are graphically captured by Matthew Sparke (2013) in his insightful analysis of the interplay between geopolitics and geoeconomics (Chaturvedi and Doyle 2010). In his view, despite the growing appeal of the geoeconomic arguments, 'their globalist vision remains deeply entangled with nationalist geopolitics at the same time' (Sparke 2013: 295). He continues:

> we cannot look at contemporary geoeconomic maps of the world without noticing their entanglement with geopolitical ideas and engagements. We cannot pretend that we have entered some sort of post-geopolitical era. Instead, geoeconomic perspectives ranging from Luttwak's new grammar to the Pentagon's discourse of disconnection defines danger as existing in uneasy tension with ongoing geopolitical assertions about national interests, national homelands, and national security. Instead, just as the early efforts to articulate geoeconomic outlooks at the start of the twentieth century were attended by some of the starkest geopolitical expressions of nationalism the world has ever seen, today's proponents of geoeconomics also frequently betray their own geopolitical interests. After all, Luttwak conceptualized his own grammar primarily for the purposes of national state-craft, and it is geopolitical leaders that he therefore calls upon to protect 'vital economic interests by geo-economic defenses, geo-economic offenses, geo-economic diplomacy, and geo-economic intelligence'.

Whilst acknowledging the primacy of nation-states, another way in which place and space are now celebrated within a more globalized world order is the re-emergence of geoeconomic regionalisms. Regions, regionalisms, and regionalizations are, at once, a result of globalization, and a challenge to it. Regions have long been part of the international system (and non-systems), but it must be acknowledged that in this progressively more translateralist world, the concept and symbol of 'region' has been increasingly used to challenge the power and existence of any uniform model of macro-geopolitics.

Nation-states remain the most vociferous players, each one jostling from different and often ambiguous positions, as to which countries should be included within and without these new regional cartographies (Nieuwenhuis 2013). In a book currently under production, (Doyle et al. 2016), we list the properties of these emergent positions with special reference to the US, India, Japan, China and Australia. Due to the nascent nature of these regional constructions, it must be stressed at the outset that these positions are not universally adopted by particular nation-states. In some countries, it may be a particular think-tank, a specific political party, or one branch of the governmental bureaucracy which advocates a version of 'Indo-Pacific' regionalism. This fits in nicely with Jayasuriya's concept of a regulatory regionalism (Jayasuriya 2008): that certain constructions are championed by actors, networks, and specific bureaucratic and epistemic communities within and outside of governments, rather than being examples of a whole-of-government understanding.

Despite this lack of coherence, however, there is no doubt that amongst think tanks, national policy-making and epistemic communities, these concepts are in their ascendancy, and on occasions, certain positions become dominant.

These attempts at regional framing are usually lacking in real intent to develop pan-regional co-operative identities, securities or economies (though these exist as subservient traditions). Rather these regional conceptual maps are usually understood by regionally situated nation-states as depicting new bi- and multi-lateral relationships and allegiances designed to draw lines *in the sea* denoting 'natural' borders and 'no-go zones', past which other specific nation-states are either admitted or excluded 'entry'.

In the case of the United States, a super-region like the Indo-Pacific is sometimes also imagined in exactly the opposite way – as a *non-region*, or *continuum* which (at least in narrative terms) denies the existence of the formal politics and histories drawn on maps by nation-states. Instead, a super-region demands free and smooth movement through time and place – national and regional borders are by-passed or passed through, as the continuum is a globally referent object of security. Instead of constantly protecting and securitizing regions and borders, it is now flows, routes, and sealanes which become important in efforts to secure geo-economic spaces in a more fluid and time-specific manner (Ryan 2013). Whereas traditional nation-statist regions were more perennial, this form of securitizing space as *non-region* is more intermittent, more temporal.

The United States has been undergoing a reassessment of the strategic importance of the Indian Ocean Region in recent years due, in part, to the growth of a range of non-traditional threats, and since the growing economic and military importance of both China and India which challenge US dominance in the region. Indeed, it has been asserted that, 'the Indian Ocean may be the essential place to contemplate the future of US power' (Kaplan 2010: xiv), and that, 'Only by seeking at every opportunity to identify its struggles with those of the larger

Indian Ocean world can American power finally be preserved' (Kaplan 2010: 323).

It is claimed that the 60th anniversary Ausmin meeting in San Francisco in September 2011 marked the 'pivot point' at which both Australia and the United States began to 'redefine their region not as the Asia-Pacific, but as the Indo-Pacific' (Sheridan 2011). Buzan has recently argued that these regionalized reactions to the ongoing rise of China have led to the generation of 'a weak but definite Asian supercomplex'. This trend, he suggests, is being reinforced both by China's turn to a harder line policy since 2008, and by an increase in the strength of United States regional linkages as part of its role as an intervening external power in South and East Asia (Buzan 2012). Buzan goes on to state that:

> The idea of an 'Indo-Pacific region' sometimes mooted by the Obama administration, is so vast as to make a nonsense of the concept of 'region' ... this fits with a longstanding and very clever anti-regional diplomatic tactic of the US ... By defining itself as part of various super-regions (the Atlantic, Asia-Pacific, the Americas) the US both legitimizes its intrusions into them and gives itself leverage against the formation of regional groups that exclude it (respectively Europe, East Asia, Latin America). This pattern is repeated if one looks more narrowly at strategic balancing behavior, the two being related aspects of the Asian supercomplex. (Buzan 2012)

As discussed earlier in this chapter, the Indo-Pacific, when viewed as a super-region, 'maritime commons' or 'the continuum' allows a global power 'access to all areas', at specific junctures in time. By invoking the existence of such a super-region, or non-region, it actually denies the existence of regions as drawn on maps, or forms of regionalism which are essentialist in terms of hard-and-fast borders (as some countries, some peoples physically live near others). The US continuum denies any 'special relationship' forged between neighboring peoples. Indeed, if such a special relationship should prove an impediment to the smooth flow of trade, or the exchange of information necessary to secure the homeland, then it will be dispersed. This is, of course, the key difference between a vision of the Indo-Pacific from the point-of-view of a super-power, and its construction under the gaze of smaller or middle powers such as India or Japan (interestingly, the Japanese concept of the Indo-Pacific usually includes the coast of Southern and Eastern Africa, and the west Indian Ocean, whereas the US and

Indian constructions usually end at the perimeters of the USPACOM command-structure zone, to the immediate west of India). Admiral Samuel J. Locklear, C-in-C, PACOM, would view Indo-Pacific more in terms of the continuum of security spectrum and invoke the more fluid geographies of flows.

> Whether the name is Indo-Pacific or something else, when I am sitting in my office looking at a pretty detailed chart of my entire jurisdiction, I *view it as a continuum of security requirements, not broken down by historical perspectives of the different oceans*. We *think 'one continuum' is a good concept*. However, it's not just about the Indian Ocean. It's about the connectivity of these large economies, the large core populations, and how things have to move. Take that to the next level and you have the cyber commons and the space commons. Ships and airplanes travelling across the Indian Ocean, whether it be to the Arabian Gulf or through the Straits of Malacca, are critical for trade and flow of energy sources. The PACOM helps protect these routes. (cited in DeSilva-Ranasinghe 2013; emphasis added).

The notion of the declining power of the United States (or its depleted hegemonic reach) needs to be further developed here. In attempting to hold its hegemonic position on a more limited budget, within a very nuanced and ever-changing series of geostrategic games, the US has had to respond in a number of ways using *austerity* as a security mantra. Despite the existence of a still large expenditure, defense spending has declined from approximately 60 per cent to 20 per cent of total US resources. These reductions in defense spending have been further exacerbated by the GFC in 2008–2009. In short, the US can no longer exclusively play (or afford) traditional geostrategic games. Whereas, in the Cold War, the lines were usually visible (sometimes manifesting themselves literally as walls dividing 'us' from 'them'), this new geostrategic game is more amorphous, ultimately flexible, temporal and many-sided. It is geo-strategic string theory. Desperate to uncover and disavow the secrets of global black holes, it sees geo-political space in ten dimensions, not the usual four (three spatial, and one temporal). The US has deterritorialized (and, when it suits the perpetrators, post-politicized) more permanent boundaries *outside* its borders (whereas its homeland boundaries and borders are further strengthened, politicized and securitized in a more realist sense).

There are a number of other geopolitical narratives from the past which ring true here. First of all, in some ways, in order to protect its commerce on seaways, landways and airspace, the military now resembles the anti-bellum cavalry protecting the migration routes of the early white settlers moving across the plains of the Homeland from East to West. And these western metaphors ring true when we imagine this Indo-Pacific space as yet another 'push to the west' (beyond California). The US navy and air force are, in effect, 'riding with the wagons' to protect them from the savagery of 'the natives' who have already been stripped of their livelihoods. At this juncture in global colonialism, however, no treaty will be necessary as sovereign space does not exist outside the Homeland, the national borders of temporary 'burden sharers', and the drifting trade flows they protect. This is invasion of the Commons using Neo-Terra Nullius (or, in this case, Terra Mare) as its doctrine.

Next, as the US is declining in real terms in relation to both its financial and military might, it seems to be deploying strategies and tactics used for centuries by comparatively powerless minorities. To fight Goliath (China in another life), the old ways of land armies with distinct battle-lines can only mean annihilation for the US. The move to air and the sea are, in part, a response to this. By having the ability to enter the fray at will by securing their markets and trade routes at certain times of passage (and then retreating), and by virtue of their ability to define where the fray is, the military – often in the form of 'Navy Seals' (Ryan 2013) – becomes more guerilla-like, more terrorist-like, in their activities. They become reminiscent of the Naxalites in South Asian forests, guarding their roads as their enemies enter their domain, striking in relative darkness, and then retreating.

Finally, in many ways, (and sometimes paradoxically) The US's strategic form can be viewed as more *post-political* (at least pertaining to its dominant narrative). As the US power recedes in relation to the power of China, it can be understood to be deploying strategies and tactics more reminiscent of new social movements (NSMs) than nation-states. Characteristics of NSMs are many, but some relate to a form of politics which allows social movements to cross beneath, around and over more traditional nation-statist borders. Social movements play a politics which is at once local and global; social movements play the long game, not the short game; social movements are multi-headed, and their goals are ambiguous, ever-shifting. The temporal nature of goal-oriented networks with broader social movements allows strange bedfellows to make alliances, as the territories and issues are shifting constantly. So, as we will argue later, apart from the military taking the issue of climate change from the global environment movement (after years of rejecting it), it may also have adopted some of its organizational structures and strategies, which allowed the movement to sustain its moral force as one of the leading social and political movements across the globe for 40 years or more.

These transient geo-strategic partners and bedfellows (bilateral monogamy seems to be a thing of the past) provide significant 'burden-sharing' (and cost savings) advantages. New regional constructions such as the IPR provide promiscuous forms of co-operative security such as that shared between India, Japan and Australia. In an article entitled 'Building Bridges over the Sea', Tetsuo Kotani (working for Japan's premier security think-tank, the Japan Institute of International Affairs) reinforces Prime Minister Abe's 'democratic security diamond', 'as a key enabler for good order at sea' (Kotani 2014). Furthermore, this vision, ably supported during three meetings of the Japan-India-US Trilateral Strategic Dialogue on Security Issues in the Indo-Pacific Region (held between November 2011 until March 2013), includes the Indo-Pacific geopolitical/geoeconomic zone as a construction which 'maintain(s) a liberal rule-based maritime order'.

In fact, this version of the Indo-Pacific is more reminiscent of a neo-liberal and neo-securitized dis/order. As mentioned at the outset of this chapter, although neo-liberalism champions movements of resources, trade and finance across borders, it only does so when this 'free movement' does not impact upon the interests of the powerful – so too with the case of neo-securitization.

Climate as omniscient neo-security

As the Pentagon loses its outward control, its *inward* power is being heightened. In an article entitled 'The Amazing Expanding Pentagon', Cambanis (2012) details the manner in which the Pentagon has captured US foreign policy through 'mission creep'. Cambanis writes:

> What 'military' means has changed sharply as the Pentagon has acquired an immense range of new expertise. What began as the world's most lethal strike force has grown into something much more wide-ranging and influential. Today, the Pentagon is the chief agent of nearly all American foreign policy, and a major player in domestic policy. As well its planning staff is charting approaches not only toward China but toward Latin America, Africa, and much of the Middle East. It's in part a development agency, and in part a

> diplomatic one, providing America's main avenue of contact with Africa and with pivotal oil-producing regimes. It has convened battalions of agriculture specialists, development experts, and economic analysts that dwarf the resources at the disposal of USAID or the State Department ... The world's most high-tech navy runs counter-piracy missions off the coast of Somalia, essentially serving as a taxpayer-funded security force for private shipping companies ... Super-empowered and quickly deployable, the Pentagon has become a one-stop shop for any policy objective, no matter how far removed from traditional warfare. (Cambanis 2012)

Climate change is just one of these non-traditional policy/security areas under the increasing influence of the military. As the US military learns from, heeds and utilizes the strategic lessons and structural forms of new social movements (like the environment movement), it has also co-opted the language and moral superiority of environmental movements. Particularly, the politics of climate change have now been increasingly used by militaries around the world to broaden their policy reach.

Cambanis contends that the Pentagon has emerged as a progressive voice in energy policy, endorsing climate change and financing research into renewable energy sources. He goes on to state:

> With little fanfare, the Pentagon – currently the greatest single consumer of fossil fuels in all of America, accounting for 1 per cent of all use – has begun promoting fuel efficiency and alternate energy sources through its Office of Operation Energy Plans and Programs. Using its gargantuan research and development budget, and its market-making purchasing power, the Defense Department has demanded more efficient motors and batteries. Its approach amounts to a major official policy shift and huge national investment in green energy, sidestepping the ideological debate that would likely hamstring any comparable effort in Congress. (ibid.)

This move towards more efficiency (and an increasing interest in extra-carbon technologies) is of particular interest here. In this manner, climate change moves from just a *threat-multiplier* to a *force multiplier*, allowing the US military to operate in 'external fields' more efficiently, for longer periods away from the Homeland. This is critical as conflict points are increasingly unpredictable as to their location, existing outside from more costly, more permanent bases. The adoption of 'greener', more mobile 'combat HQs' allows force to be deployed for longer, and more flexibly.

In an article which largely focuses on the Australian military's response to green initiatives, Press et al. list some of the green combat initiatives deployed by North American defense forces:

> The US Army is developing an electric vehicle fleet in order to reduce its reliance on fuel on the battlefield. The US Department of Defense is increasingly turning to microgrids to ensure self-contained energy generation and assuredness during critical operations. There are 454 renewable energy initiatives currently underway or under development by the Department. The development of more efficient and longer lasting batteries and fuel cells to provide portable power systems for troops is a US defence priority. In 2010, the US Air Force conducted the first successful test flight of an aircraft powered by a biofuel blend. It aims to use alternative fuels for 50% of its domestic needs by 2016. Increased use of alternative fuels in US tactical fleets and systems is an important consideration for ADF capability planners seeking fuel interoperability between national platforms. (Press et al. 2013: 27)

Also, in a report released by the US Department of Defense in 2011, commonly referred to as the 'Pew Project on National Security, Energy and Climate', arguments pertaining to the morality and rhetoric of cleaner carbon futures, coupled with force multipliers, were both evident (Schario and Pao 2011):

> The U.S. Department of Defense (DoD) is accelerating clean energy innovations in an effort to reduce risks to America's military, enhance energy security and save money, according to a report released today by The Pew Charitable Trusts. 'From Barracks to the Battlefield: Clean Energy Innovation and America's Armed Forces' finds that DoD clean energy investments increased 200 per cent between 2006 and 2009, from $400 million to $1.2 billion, and are projected to eclipse $10 billion annually by 2030.
>
> 'As one of the largest energy consumers in the world, the Department of Defense has the ability to help shape America's energy future,' said Phyllis Cuttino, director of the Pew Clean Energy Program. 'DoD's efforts to harness clean energy will save lives, save money and enhance the nation's energy and economic future. Their work is also helping to spur the growth of the clean energy economy.'

The work of Emily Gilbert is particularly salient here. She argues the US 2010 Quadrennial Defense Review builds on the previously mentioned Center for Naval Analyses (CNA) report of 2007, which connects climate change with failed states, humanitarian aid, terrorism and mass migration scenarios (Gilbert 2012: 2). She critiques this trend for several reasons. Firstly, the military takes a narrow, traditional view of security, as we've already described above. This is based on nationalistic, defensive, territorial lines, viewed in statist terms. It is furthermore a model of external threats, based on the idea of resource conflict, which,

> coheres easily with the competitive frame that has been established between China and the US, as they vie not only for economic ascendency and resource-acquisition, but also for energy security and environmental policies and initiatives. In this vein, Thomas Friedman has proposed a militant green nationalism, something along the lines of a triumphalist Green New Deal that will recapture US global hegemony. (Friedman 2009 quoted by Gilbert 2012: 3)

In this manner, the military is further legitimized, 'to the detriment of formal and informal politics' (Gilbert 2012: 4). In their expanded roles as providers of disaster and humanitarian relief, they are given entrée to, or encroach upon, the roles of civilian development and aid: 'This is part of a worrisome trend of the rise of an "aid-military complex" and military "encroachment" on civilian-sponsored development' (Hartmann 2010: 240). Furthermore, the militarization of the phenomenon does not address the causes of climate and environmental insecurities in any way. There is never any discussion of the fact that the vast majority of the earth's resources are consumed by an ever-decreasing few. It merely entrenches the role of the military in defining the problem – in the worst-possible-outcome/worst case scenario sort of way, and this becomes 'the basis for actions in the present' (Gilbert 2012: 4).

Her second major concern is that the environment is being mobilized and cast as the enemy. Again, this reality has already been alluded to in this chapter's opening pages, but Gilbert provides a useful addition to this discussion, as she argues that this has the useful effect of resurrecting and perverting a view of 'the commons' which now actually serves to defend national interests, as opposed to genuinely common ones:

> Either way, nature is an externality to be managed as the resurrection of the concept of 'the commons' in these debates affirms

> (see Posen 2003). Advocacy groups and government representatives alike are using the 'commons' to inform their perspectives on climate change security. Abraham Denmark and James Mulvenon explicitly delineate the concept's legacy to Garrett Hardin's controversial piece, 'The tragedy of the commons,' and his argument that 'Freedom in a commons brings ruins to all' (Denmark and Mulvenon 2010: 7–8). Rather than privatization, the contemporary version of the polemic posits that military force is necessary to prevent the misuse and abuse of navigable passageways. (Gilbert 2012: 5)

Gilbert refers to the 'complex web of collaborations' addressing climate change, which she describes as a 'military-industrial-academic-scientific complex' (2012: 6). Transformative technological innovations are presented as being of immense social benefit, and hence, it becomes easier to justify enormous amounts of money being transferred to the military and its privatized civilian partners to work on these carbon friendly technologies. The problem, of course, of funneling resources (however green) through the military is that resources are drained from other sectors, 'unless they are working in partnership with the military' (2012: 8). Meanwhile, returned military personnel are reintegrated into civilian life through various 'green' mechanisms, like 'green training' and 'green jobs' initiatives. This is just another way of legitimating spending on the military:

> Domestic measures to address energy security are put forward as calculable, rational and even compassionate measures, while the 'foreign' threat is presented as non-state, elusive, and undetermined – and hence coherent with much of the discourse around diffuse 'new wars' and terrorist threats (Kaldor, 2006). At the same time, there is also greater convergence between the inside and the outside, and between the environment and the military in the ways that the discourses are mobilized and mapped out (Cooper, 2006). Indeed, as Mikkel Vedby Rasmussen notes, there is a coherence between pre-emptive military doctrines and precautionary environmental strategies: both are based upon a rationale for urgent action based on anticipated future disaster scenarios (Rasmussen 2006: 124). Notably, however, it is only when environmental issues are harnessed to security claims that the precautionary approach gains traction. (Gilbert 2012: 10)

Ferguson would argue that this is typical of militarism's 'double move': on one side of the coin, war is constructed as being 'over there'; whilst on another side, the 'second move saturates our daily lives with warness' (Ferguson 2009: 478).

Conclusion

The environmental and climate change movements have been captured by the military. Not only has the grand narrative of climate change been co-opted, warped and re-routed by the proponents of *green security*, the very forms of new social movement resistance have been copied and reworked to suit these most recent geopolitical moments. In these multi-layered, multi-directional spaces, neo-liberal economics and neo-securities are one. At this chapter's outset, we described the forceful emergence of neo-liberalism as an economic doctrine in the late 1970s and 80s. By the time of the first Earth Summit (1992), narratives of security had also begun to weave their way into environmental doctrines in much the same way as economics had done in the decade earlier. Perhaps we have to question what new social movements are? Perhaps there is as much a Security Movement as there is a Neo-Liberal Movement, as there is an Environmental Movement. All these movements are, in their current form, off-springs of capitalism. The widely-held concept that 'true social movements' emerge from the bottom-up (are grass-roots by definition), and radically challenge the status quo (Doherty 2002), needs to be questioned. Perhaps it is the 21st Century's Neo-Security Movement (NSM) which is, in fact, the most powerful new social movement (NSM) on the planet. Like the environmental movement, and the neo-liberal movement which followed it, the neo-security movement has moved to define every part of our lives, in both the public and private spheres, inside and outside the state.

Perhaps the NSM is the ultimate NSM. But NSM theory is only appropriate in post-political societies, sometimes found in the global North, where capitalism 'has won'. In most parts of the Indo-Pacific, where people battle for daily survival, the language of post-politics actually strips away the importance of their place – and that place determines how they will live, how long they will live for, and how they will die (Catney and Doyle 2011). In this construction of omniscient climate security, there is no distinction between luxury emissions and survival emissions. They have been washed away. The politics of repression has been stripped out of Indo-Pacific geo-economic equations.

Climate change in the continuum constructs global spaces – in this case the Indo-Pacific – as connected threads of gold through lawless darknesses; as networks, pathways and trading songlines through black waters and evil airs held together by strings of liberal values. The depiction of nature is still a realist one – the essential nature of nature is a maelstrom, is still anarchical – and nation-states (at least the good and the true ones) must order it, and call its marauding tribes to account.

This is omniscient security. It secures the earth using a climate narrative which cannot be seen, smelt or touched in the lifetimes of humans. It strikes everywhere and nowhere at once. It is arbitrary and can be deployed only by the most powerful and their piecemeal coalitions to protect the 'free flow' of capital. It is both a conflict and force multiplier. This latter interpretation allows the US military and its temporal allies to strike harder for longer (through greener, non-carbon energy intensive warfare) when fighting in intermittent conflict spaces away from more ephemeral combat bases and headquarters and, of course, the Homeland. In this vein, the Indo-Pacific is now largely understood (in US defense terms) as a series of geoeconomic routes and elastic zones which need to be secured; sovereignty (at least in an external affairs sense) becoming something amorphous and arbitrary (though no less powerful); and climate change is used in a manner which accentuates this idea of 'comprehensive security' in a 'continuum' or super (non-region), sometimes called the Indo-Pacific. Accordingly, whilst advocating climate crisis imperatives, the military can now make preemptive strikes, as it is the only one which has 'the lift capacity' to secure trade-flows, 'protect' the most 'climate vulnerable' by 'swarming' to secure 'storm surges', and utilize both armed and unarmed 'intervention relief' in order to establish 'neo-Malthusian anticipatory regimes' (Adams, Murphy and Clarke 2009). As a consequence, Indo-Pacific citizens and their leadership are now being made markedly insecure, disciplined by predominantly foreign geoeconomic forces.

This is simply plutocracy – the rule of a wealthy global elite – desecuritizing and disciplining weaker domestic economies and democracies. This is a very different kind of 'foreign policy'. In its place, there is a *re*-securitization; but one which only further secures the power of the elite to the detriment of the many. New geo-economic lines, new geopolitical maps are drawn marking new borders and boundaries between the haves and the have-nots, crossing and then erasing the histories, cultures and politics of troublesome nation-states and non-states. This is a new type of colonization: a transition from state-led to TNC-led colonization, ably supported and protected by powerful (but at the same time declining) military regimes such as the United States, the very same country which refused to sign the Kyoto Protocol as it cut too deeply into its cherished notion of freedom.

7
Climate Justice: An Attempt at an Emancipatory Politics of Climate Change

In recent times, the greater prominence of climate discourses amongst majority world environmentalists has occurred due to the fact that some of the world's biggest polluters and/or reliers on fossil fuels have still not signed or endorsed the climate change protocols in Copenhagen, Kyoto, Johannesburg, Bali, and others in any realistic fashion. As far back as October 2002, for example, 5,000 people from communities in India, including international NGOs, gathered in a Rally for Climate Justice in New Delhi. This rally was organized to coincide with the United Nations meeting on climate change (Conference of Parties 8 – COP8), and was organized by the India Climate Justice Forum, including the National Alliance of People's Movements, the National Fishworkers' Forum, the Third World Network, and CorpWatch. At this 2002 protest, Friends of the Earth International (FoEI) expressed frustration with climate change negotiations:

> But climate negotiations show no progress and communities are calling for urgent action to address climate change and to protect their livelihoods in a manner that is consistent with human rights, worker's rights, and environmental justice ... Given the entrenched opposition to action from the fossil fuel industry and governments like the US and Saudi Arabia, environmental organisations joined forces with social movements in order to progress this most urgent agenda. The window of opportunity to prevent dangerous climate change is closing fast and, for many communities, the impacts are already alarmingly present.

Also, further majority world acceptance of the climate agenda has emerged due to the fact that Northern climate rhetoricists have successfully focused on poor island states – as well as less affluent dwellers and coastal fishworkers on coastlines of the Global South – who will be the principal victims in global climate change scenarios. In this manner, climate change has metamorphosed from a purely elite, scientific, neo-liberal northern issue into one which can usefully fit into the environmental justice agenda of the South, but not without evoking in the first place serious dilemmas both before the governments and the civil society.

But of course, the poor and marginalized are always on the periphery. They are there today, suffering most as part of a global food crisis; they are there today, experiencing the worst of a Northern-induced global financial crisis. Of course, they are always most vulnerable, whatever the geopolitical context, or the issues which define it. It is rather trite, therefore, to point out their vulnerability in some future crisis which might happen. Rather, it is more likely a strategic means utilized by the neo-liberal North, a way of garnering support from those on the 'Global Left' who are similarly bewitched by the glorious power of the climate symbol. As pointed out by O'Brien and Leichenko despite widespread recognition that there will be 'winners' and 'losers' with both climate change and globalization, the two 'global' issues are rarely examined together. They introduce the concept of 'double exposure' as a framework for examining the simultaneous impacts of climate change and corporate globalization. The term refers to how certain regions, sectors, ecosystems and social groups will be confronted both by the impacts of climate change, and by the consequences of globalization, resulting in new sets of winners and losers.

The key questions which haunt this chapter, then, are as follows: Can the climate discourse be re-configured, re-reterritorialized to provide a place where issues of environmental justice and sovereignty are paramount, rather than neo-liberal responses to climate? Can climate change give a voice to the global periphery? Can it be used as a vehicle for emancipation? How have the ideas informing climate change been utilized by actors in the global South, and what have been their repertoires for action?

Climate change and 'the left'

In Chapter 4 of this work, we argued that the roots of climate change – in an ideological sense – lie with the political Right. In more recent times, as touched on above, more emancipatory groups, from both the global South and the more progressive North have attempted to use elements of more Leftist ideology to invade climate narratives, co-opting them for their own political projects. Before we head to our case-studies, some key elements of Leftist ideology must be briefly explored as they relate to climate change. Obviously, there are more 'critical' theories that question the established ways of the world and how it is organized. These critical, more 'collectivist' perspectives move the focus of analysis away from both individuals and states (O'Brien and Williams 2010), and emerged largely in direct opposition to right-wing thought, be it liberal, neo-liberal or conservative.

Two particular collections of theories have dominated leftist responses in relation to climate change: socialism and anarchism. For both, capitalism is a major problem and, as it currently stands, it cannot be 'greened' in order to produce a 'carbon-friendly' world.

Market relations are seen as inherently and ultimately exploitative and unequal, with nation-states and corporations as generally representing the interests of dominant classes (O'Brien and Williams 2010).

The eco-socialists (or socialist ecologists/environmentalists) attempt to mesh ecological principles with those of another, more traditional set of political theories revolving around Marxism. In this instance, its ecotopian visions very closely match that of certain types of socialism. Socialist ecologists are anthropocentric enough to believe that they should move strategically from issues of social justice to ecology, not vice versa.

This political philosophy generates its own brand of green economic response to climate change. In a more centralized green economy, the state would not only monitor and legislate against the excesses of climate-degrading companies, but the state would actually drive a green, non- or low carbon economy by being the major shareholder in it. The issue here is not whether there are inadequate resources to go around but, that under capitalism, these resources have been wrongly concentrated in the hands of the few. Under socialism, the people, as a collective, would actually own their resources. Consequently, green socialists do not advocate the existence of population 'carrying capacity' or 'resources scarcity': poverty, then, is seen as a result of capitalism, not the fault of poor people producing too many children.

The biofuels debate is of interest here. Where alternative sources of non-carbon based energy are enthusiastically embraced by neo-liberal greens, biofuels solutions are heavily criticized by climate change advocates on the left, as these agricultural monocultures denude landscapes, stripping valuable food-bearing farmland away from food production. In this manner, biofuels are seen as major contributors to the pre-existing global food crisis, regardless of the 'climate saving' merits.

In the eco-anarchist position, *climate authority* – through the auspices of *expert knowledge* – is not developed through necessity to protect people and the earth from themselves; but rather as a means of the powerful to promote their own interests, and to subjugate the interests of others. The anarchist tradition is often hostile, for different reasons, to the state, liberals and Marxists, stressing instead the importance of non-hierarchical human social organization, decentralization, and self-government. In an anarchist society, environmental and climate change problems can only be solved with maximum local autonomy, as society is seen as connected rather than severed from the ecosystem within which it resides.

Like socialists, most anarchists believe an ecologically viable society is incompatible with capitalism and its need for continually expanding markets and the built-in obsolescence of consumer goods. However, rather than wait for the downfall of the capitalist system, they seek to build more sustainable economic systems from below. Bioregionalism is one example of this green anarchist response. Bioregions are naturally occurring boundaries which define ecological and social communities. Bioregionalists also argue for the importance of *place*. The concept of place sees individual humans as necessarily having both a physical and spiritual attachment to both a community of other people, as well as to a bio-physical, defined space. They reject the new-found extreme mobility of advanced industrialized societies. In this manner, green anarchist economics rejects the globalization of economic systems with its attendant reliance on free markets and capital flight.[1]

The emphasis on the importance of place and local autonomy has led to a burgeoning of the community garden movement in some first world cultures (Nettle 2014). It has underpinned the movement by citizens' groups to reclaim common space for food production and community events, enabling communities to become more self-reliant and climate friendly, better able to preserve and grow cultural foods, more rooted in the place they live, while reducing the global climate impact of their 'food miles'. In the first world, the use of alternative technology has also arguably been inspired by social ecology theory and its call for emancipatory, collectively-owned technology. We should note, for example, the trend towards local community 'green' electricity generation such as in Denmark and the Netherlands, where 50 per cent of energy generation is now decentralized (both 'green' and fossil fuel based). In 2006, Britain's May 2006 budget included an allocation of £50 million to support community-based 'micro-generation' initiatives (Catchlove 2006). Other grassroots community developments such as city eco-village housing projects, community environment parks, and certain greener city initiatives are also modeled along social ecology lines.

Yet for many communities around the world, particularly in majority world countries and for the peoples of First Nations, this is simply the way things have always been done. Furthermore, it is important to understand that the ways in which third world cultures have traditionally produced food and organized their economic realities within decentralized polities is now also sometimes under threat in this era of globalized production.

For the remainder of this chapter, we will look at specific case-studies taken from Labour, Religious and Green Movements.

Unions and religious groups

The geopolitical position of particular emancipatory groups usually – but not always – is critical in determining both the manner in which climate 'ideas' are expressed, and the political pathways which are pursued. Most northern groups, whether they be unions, peoples movements, church or green groups, with important exceptions, usually espouse ideas and repertoires which are not particularly at odds with advanced industrialism and global capitalism. Of course there are regional differences also, but those groups occupying the Southern peripheries are more likely to critique capitalism per se, as the main culprit contributing to or causing climate crisis.

The union movement is a wonderful example of this point. Climate change, and international trade union (ITU) responses to it have not been deeply studied but, in an excellent article on trade unions (2013), Felli argues,

> it shouldn't come as a surprise that, over the last few years, ITUs have produced various documents and resolutions, organised conferences and written reports, dedicated resources (material and human) and taken part in international negotiations on climate change. These elements seem to point to the fact that trade unions, at least at the international level, are increasingly concerned with the issue of climate change. (2013: 3)

Felli states that ITUs basically share a broad position on climate change, despite their many differences. They take the climate science as fact, and aim to combat climate by means of sustainable development and 'just transition', which basically means that workers should not pay the price for social and economic adjustments necessary to deal with climate change. There are differences between ITUs over the course of action to be taken specifically, there are, different strategies or alliances, but not the existence of climate change and the need to deal with it. He argues the dominant ITU approach is sometimes anti-neo-liberal in its policy direction (at least in its rhetoric) (2013: 7). ITUs are only committed, however, to technological fixes to the climate crisis and to the notion of green capitalist economy and green jobs, and therefore they fall in line broadly with 'ecological modernization' discourses, as discussed in Chapter 2 (2013: 8). In this way, they are supported by the International Labour Organization and the United Nation Environment Program (UNEP), which show almost identical commitment to these principles. ITUs argue for a shift to a green economy which needs to be supported by governmental policy and investment:

> This very broad notion of a green Keynesianism framed in an ecological modernisation perspective, and coupled with demands for a 'just transition', is the overlapping strategic consensus of the international trade union movement. Unsurprisingly, it is a consensus which deals essentially with the processes of value redistribution, but which, apart from the promotion of 'green' sectors, has little to say on the very social relations of property and production that are at the heart of the growth imperative and the climate crisis. In this sense, we are not dealing at this level with a strategy aimed at changing the context or the broader political economy. (Felli 2013)

Felli goes on to identify three main union strategies on climate change:

1. *The Deliberative Strategy*: where he places the ITUC, ILO, UNEP etc. in a non-conflictual, collaborative approach which, while sometimes rejecting neo-liberalism, still ultimately accepts many of its policies and market-based approaches. It is broadly cosmopolitan without challenging existing social and productive relations.
2. *The Collaborative Growth Strategy*: this is likely to include the more traditionally unionized workers in industries most likely to be affected by changes in the economic system toward 'green' production. He places the International Federation of Chemical, Energy, Mine and General Workers' Union (ICEM) in this category. Felli argues that these unions are strongly oriented towards economic growth, and finds that this approach incorporates elements of 'weak'

ecological modernization. This strategy has a strong bias towards Northern trade union interests, and includes Australian trade unions in this categorization (2013: 16). He writes: 'this strategy seeks to alter the balance of power within nationally-based social formations. It can, therefore, be said to be a context-changing strategy in political-economic terms without, however, questioning the broader context of capitalist accumulation within fragmented national territories' (2013: 17).

3. *The Socialist Strategy*: here, for example, he includes the International Transport Federation, which critiques the growth imperative and international capitalism.

   As Felli suggests, the Australian example has been singled out as almost an ideal type of the second strategic pathway. An important example to consider here is a deal struck between the Australian Conservation Council (ACF), The Australian Council of Trade Unions, the Property Council, the Institute of Superannuation Trustees, and others, which has only led to a 'light green' approach – not surprising due to the ACF's and the ACTU's overly close relationships with the then ruling Labor Party (Goods 2011). Ariel Salleh has been scathing in her criticism of this form of climate change policy. She writes of the ACF's climate change credentials as follows:

> The Joint Statement advocates a new economic sector of green industries for the manufacture of globally competitive product innovations and services. This promises 500,000 green jobs, with re-skilling for Australian trades men and women. There is no engagement with the grassroots movement call to reconfigure 'the social contract' and no sense of Australians as ecological citizens with responsibilities that are global in reach. The technocratic focus also marks UNEP's *Global Green New Deal* (2008), which is essentially a 'development' model where people become 'human capital' and their habitat is quantified as 'natural capital'. By this reckoning, common land, water, biodiversity, labour, and loving relationships are pulled away from an autonomous web of eco-sufficiency. (Salleh 2012: 121)

This third approach touched upon by Felli is more likely to be encountered in majority world situations. The peak body of South African Trade unions, the Congress of South African Trade Unions (COSATU), takes an overtly 'climate justice' approach to the issue of climate change. For example, in a 'call to action' policy document entitled (2011) 'A Just Transition to a Low-Carbon and Climate Resilient Economy', they argue the following:

> Climate change is caused by our present system of production, distribution and consumption, a system which is both unjust and unsustainable. We have to change our way of producing energy, the way we work, produce goods and provide services. We have to create a low carbon economy in order to preserve our planet for future generations and in order to reduce the impact of climate change on water, food, livelihoods and other necessities. COSATU is committed to making a just transition to such an economy. This means putting the needs of working and poor people first in the social and economic changes ahead of us' (COSATU 2011: 1) ... The climate change challenge is not the only crisis we face in South Africa. The climate crisis is linked to the environmental, unemployment, poverty, inequality, and food crises. All of these crises link back to the central economic crisis of capitalism (ibid.: 68).

COSATU's primary concerns, therefore, deal with the broad issues of water security, food sovereignty, land reform, livelihoods and gender issues. They argue that these are central to dealing with climate change.

This ideological and repertorial split between North/South is also very visible in many institutionalized religious associations. As mentioned in Chapter 2, conservative concepts of the Flood have been used by an array of diverse religions and cultures to throw their support behind climate change reforms. Operation Noah, formed in Britain in 2000, after experiencing severe storms and flooding (Bodenham 2005: 113), is a good example of this ideological vision being put into practice at the grassroots level. For example, in Spring 2004, an exploratory interfaith project was brought together by Nottingham Interfaith Council, involving young people from the Jewish, Muslim and Christian communities in the city. With public funding, the young people employed a musician and a painter. They took part in a series of four meetings of creativity and learning the role of Noah in the Bible and the Koran, and its relevance at a time of environmental change (ibid.: 114).

Of course, religious responses to climate change include many groups who, although also using conservative language, promote themselves in direct opposition to climate change, using religious dogma to challenge the veracity of climate change. These religious groups object to

Noah's Ark being used as a means to champion the climate change debate. The routes that climate fear takes (sometimes divergent and in some cases in a collation form) are highly convoluted and thus difficult to map to perfection. The complex interplay between geopolitics and religion has been noted by a number of political geographers. One of the most intriguing routes that geopolitics of climate fear has taken, particularly in the case of the United States, is that of 'evangelical environmentalism', which claims to base itself on a Biblical worldview. What is worth noting in this case is the ways in which the fear expressed by the Other (the environmentalists – whether Christian or otherwise) is being dismissed as 'un-Christian' with the help of a counter-fear through a crafty manipulation of fear-hope interface anchored in Biblical metaphors and language. By way of illustration, the 'crusade' of *Cornwall Alliance for the Stewardship of Creation* against the so-called 'Green Dragon' is highly critical and dismissive of what they regard as 'radical' environmental movements that aim at 'saving the planet' and in the process dare to question 'Christian values that built the Western civilization' (Wanliss 2010: 16); this narrative runs as follows: the environmentalists making an appeal to humanity as a whole to make sacrifices, including 'surrender of liberties', are not only 'anti-human' and 'anti-church' but in fact are 'exploiting' religious language and sentiments in order to propagate a 'powerful rival religion, ancient as Eden, old as the ashes, against which Christians are permanently at war' (Wanliss 2010: 267). Global warming is said to be a 'dream come true for busybodies who want to micromanage the details of the lives of others' (ibid.: 236) and,

> those who chase Green peace harbor disorder in their minds because their very existence is pollution to the Earth. Their flawed theory is an alternative religion that elevates Earth and animals and creates an unquenchable dissonance in their heart. People fear because they cannot make sense of their world and are alienated from God. They feel trapped in a fatalistic death spiral from which escape is impossible. (ibid.: 262–263)

The appeal therefore becomes: 'Stop supporting the Green Dragon that wants to strike you down. Proclaim the only true hope for humankind and the entire universe – the unadulterated gospel of Jesus Christ. Resist the Green Dragon! Rise up and strike off its head' (ibid.: 19).

The preamble to 'An Evangelical Declaration on Global Warming', issued under the auspices of the Cornwall Alliance, expresses the fear that

> many of these proposed policies (on climate change) would destroy jobs and impose trillions of dollars in costs to achieve no net benefits. They could be implemented only by enormous and dangerous expansion of government control over private life. Worst of all, by raising energy prices and hindering economic development, they would slow or stop the rise of the world's poor out of poverty and so condemn millions to premature death.

Vehemently denying that the current era of global warming is unprecedented, anthropogenic and 'dangerous', the declaration says that, 'Earth and its ecosystems – created by God's intelligent design and infinite power and sustained by His faithful providence – are robust, resilient, self-regulating, and self-correcting, admirably suited for human flourishing, and displaying His glory.'

Coming simultaneously to the defense of both the fossil fuel and nuclear power industries and the poor and the downtrodden, the declaration describes 'abundant, affordable energy' as indispensable for pulling millions out of abject poverty and related risks of disease and premature death. The list of strong denials includes that 'carbon dioxide – essential to all plant growth – is a pollutant' and thus responsible for rise in global temperatures. There is also a denial of plans that promise to replace fossil and nuclear fuels, either wholly or in significant part, to provide the abundant, affordable energy needed for sustaining prosperous economies or eliminating poverty. And finally there is a denial that, 'such policies, which amount to a regressive tax, comply with the Biblical requirement of protecting the poor from harm and oppression.'

Of course, there are many more Church groups in the North of a more liberal (and even, on occasions, a more critical) persuasion. Caritas Internationalis, an organizational outcrop of the Catholic Church, although originally emerging in the first world, is now a transnational organization. Even within this one organization, there is tremendous diversity in ideology and action. In some of the first world national branches, there is an emphasis on constructing a global climate ethic – or building *capacities* – which is very much in line with the building of a 'Global Soul' as discussed in Chapter 3. In Caritas Internationalis (2009) 'Climate Justice: Seeking a Global Ethic' there is much emphasis on changing the way people *think*. Climate deniers, in this manner, are classified as wrong-thinkers, and must be re-educated in the ideas of ecological debt and intergenerational justice, and in the idea that current wealth levels in the developed world should be regarded as a 'loan' that must be repaid to the developing countries and 'the environment', in a general sense. Caritas Internationalis (2009: 4, 5) contends that people in the global North must embrace a global ethic which questions its heightened standards of living, and makes a strong call for changing models of development. It refers to the 'structural injustice' that those who have contributed least to the problem of climate change are the first to feel the effects, and that 'Unrestrained economic development is not the answer to improving the lives of the poor'. It is argued that the teaching of, and ultimate adherence to, this overarching Catholic global climate ethic will contribute to climate mitigation. A concept of 'authentic development' is pursued: one which supports 'moderation and even austerity in the use of material resources'; incorporates alternative visions of good society; demands reduction of overconsumption of resources; and the selection of appropriate technologies that benefit people and the land (Caritas Internationalis 2009: 15).

Church and other religious groups in the South seem to be less interested in mitigation and education programs which espouse alternative ethic systems, and more interested in service delivery and adaptation. Caritas Kenya, for example, though supporting the work of its central administering body in the North, is more involved in climate adaptation programs on-the-ground:

> Caritas Kenya promotes resilience in drought-prone semi-arid areas by planting drought resistant seeds that can withstand weather variations. Projects in Homa Bay are designed to combine dairy farming with bio-gas production, the residue of which is used for organic farming. (Gorman 2011)

Another example of this focus of climate adaptation is found in India's Orissa state, where Catholic Relief Services is enhancing local people's abilities to engage with climate change-induced problems:

> Catholic Relief Services is building local capacities to respond to emergencies ... of climate-related hazards by strengthening self-help groups and organising task forces to deliver first aid, plan evacuation routes and safe shelters, protect clean water sources, save grain and cash in preparation for the cyclone season, formulate sustainable crop and land use plans, and repair and construct water harvesting structures and embankments. (Gorman 2011)

The work of Caritas Internationalis is strongly reminiscent of the next case-study on transnational green group, Friends of the Earth International.

As Northern groups transnationalize, they are rapidly learning that their original northern-centric missions must be altered in terms of both ideological rhetoric and the repertoires through which these visions for change are to be articulated and put into practice. In these transnational emancipatory organizations, it is more than just 'marketing their brand' to the global South, it is also increasingly about *listening* and learning from the peoples of the Majority Worlds.

Environmental groups: the case of Friends of the Earth International

Friends of the Earth International sees itself as *not* part of the global governance state as some green organizations do, but rather, it portrays its operations more in line with a more human-inclusive, political ecological view of *environmental* and climate *justice* and *green welfare* issues. But even in such an environmentally emancipative group, there are strong pressures between its own more post-political members inhabiting the affluent, minority world, and its members in the South.

FoEI is an international confederation operating within over 70 countries. It sells a broad concept of political ecology, including strong distributive justice and welfare goals. Over recent years, it has battled to overcome intense differences between its Northern and Southern delegates (national representatives) (Doherty and Doyle 2011). Tensions have emerged, for example, relating to the overuse of English language and electronic media in the everyday running of the organization. In fact, as a result of these divisions, FoE (Ecuador) withdrew from the confederation in 2002. Since that time, there has been a sustained and committed process within FoEI to try to recognize (and, more importantly, accommodate) these differences.

In 2006, when the Federation met in Abuja to agree its first global Strategic Plan, Northern and Southern groups had vehemently different ways of constructing the political: what constituted 'communities'; what was 'democracy'; what made up 'the polis' (human and or non-human); as well as very different ideas relating to the politics of time.

Many delegates from the global South within the Confederation objected to the post-materialist concept of 'future generations', preferring to hold onto a concept of the present and the past, seeking to engage with, and resolve inequities there. Of course, in addition, in the South, interest in extending a political voice to non-humans is not high on the agenda, as the political voices of peoples in less affluent communities are not sufficiently heard. Empowerment of people, in this vein, must precede empowerment of the non-human ecosystems. Indeed, the construction of a broadened, more inclusive human/non-human polity was roundly rejected by many countries within the Confederation from the South, and there were also serious tensions between the ways in which 'communities' or 'democracy' were understood across the North/South divide. These different theoretical treatments of political forms can only be understood if we accept that the construction and understandings of 'communities' by FoE groups in the North and South are also quite different. In this vein, there are clear differences in the relationships of the delegates to their constituencies. Though all delegates at the Abuja meeting were, in most ways, political elites, the Southern delegates perceived themselves more as voice-pieces for their communities (in a traditional territorial sense), whilst the Northern delegates either saw themselves as 'the community' (in a post-political, post-territorial sense), or as 'educators' of their communities (however disparate and dispersed). Certain forms of cosmopolitanism, embedded in liberal notions of deliberative democracies and post-territories, can be sometimes more apposite in describing western European and North American delegates' notions of transnational power within the organizational network, whilst models of solidarity – more often built upon more traditional premises of 'the political' – are far more useful in understanding the politics of these FoEI groups operating in the South.

Browsing though the myriad of national position statements on climate change, it becomes apparent that this post-industrialist framing of climate change is re-enforced time and time again, particularly in European FoE groups. For example, the Belgian website advocates a 'negawatts' solution by 'reducing energy consumption, improving energy efficiency, and renewable energy development' (FoE Belgium 2010). FoE Cyprus and FoE Hungary also articulate a similar stance based on energy efficiencies. In FoE Cyprus' case, these energy savings are directed at the level of national politics, demanding 'strong national emissions reduction programs and targets' (FoE Cyprus 2010). Conversely, FoE Hungary focuses largely on the changing of individual and household behaviors:

> Hungary significantly contributes to greenhouse gas (GHG) emissions steadily increasing on the global level, leading to climate change (emissions: 150% of global average). Besides industry, agriculture, transport, households become main emitters directly or indirectly. Heating, lighting, running hot water, use of household appliances use energy primarily made by burning fossil fuels that cause significant GHG emissions. (FoE Hungary 2010)

This largely coherent European framing based on energy savings also has some significant ructions, most clearly in the case of nuclear power. In some instances, nuclear power is seen as a form of *green energy*, a lesser of two evils when compared to carbon-based energy production. FoE groups across the globe have a long tradition of being important movers and shakers in the anti-nuclear movement. Whilst FoE Cyprus actively opposes nuclear power, by rejecting 'false energy solutions that increase people's vulnerability to climate change such as oil, coal and gas, nuclear power, agro fuels and large hydropower' (FoE Cyprus 2010), FoE EWNI and FoE France – though not directly endorsing a nuclear alternative – remain quite silent on the issue.

So, understandably, outside of Europe (and to a lesser extent in North America and Australia), climate change has not always been held close by all national members of the FoE organization, particularly on the 'wrong side' of the North-South Divide. On its first emergence, there is no doubt that it was an issue which was seen as exclusively Northern, post-materialist and post-industrialist in both its creation and its relevance. In fact, regardless of changes and transmutation in framing and repertoire, this framing remains dominant. In fact the dominance of the issue – and the widespread understanding that climate change was a predominantly elitist, anglospheric pursuit – was one of the reasons FoE Ecuador pulled out of the Confederation in 2002, sparking the rash of organizational soul-searching in the middle of the decade. In many parts of the South, an industrial revolution – on the scale experienced in the North – had not occurred yet. Why then would environmentalists in the South care if western scientists and greenies in Amsterdam told them of an issue in the sky which no-one could see? Surely there were more pressing issues to be dealt with? With the North now experiencing the environmental outcomes of its own industrial excesses, why should the South restrict its peoples' own similar paths to development?

But what this campaign case study does most adeptly is to illustrate how campaign frames can change quite quickly over time and how, in these frames' transition, repertoires also alter. When climate first emerged on the international green radar, as the case of Ecuador proves, it was often rejected. In time, a number of Southern groups within the Confederation decided to move from a position of rejection to accommodation of the climate project but, in doing so, the Southern framing of the issue was also very different to that experienced in the North. As the first decade of the new century came to a close, there was increasing evidence that these more post-colonial Southern interpretations of climate change were starting to impact upon their Northern counterparts, with their largely post-materialist and post–industrialist frameworks also challenged. If anything, the climate campaign is a wonderful example to remind us that frames and repertoires metamorphose over time and space; that they are constantly negotiated and renegotiated amongst the networks of groups within regions and geopolitical groupings. There can be no doubt that vast differences in framing and actions remain the defining feature of FoE, reflecting the poles of affluence which dissect the organization. To its credit, however, there is real evidence of cross-polinization of ideas and actions from North to South, North to North, South to South, and (most critically) South to North. Indeed, this fluidity of framing and repertoire over time defines and structures this case study.

In its early days, climate – as a campaign focus – was a polarizer within this green organization, quite neatly splitting groups between North and South. The clear line between Northern support and Southern rejection of the climate agenda began to fade in the early 2000s. Now, climate's staggering breadth of ideological reach has re-mutated into versions of the climate discourse which include post-colonial and environmental justice arguments. In this light, climate justice attempts to grapple with notions of *climate debt* caused by centuries of ongoing colonialism, rather than just focusing on current *climate footprints*.

Examples of a more post-colonial climate framing are evident in the campaign position statements of FoE groups – particularly from Latin America and the Caribbean region – such as Argentina and Chile. FoE Argentina's framing clearly articulates an environmental justice position based on the concepts of climate debt and climate justice: 'We participate in the construction of a world movement in producing climate solutions that satisfy social and economic equality at the international level and the domestic level of this country – this is Climate Justice' (translated from Spanish, FoE Argentina 2010).

FoE Chile also provides a key example here. Until very recently, FoE Chile was not immersed in the climate campaign:

> Before 2008, climate change received only sporadic attention from the media in Chile, and few people beyond the technocrats that participate in UN negotiations and investors in carbon markets had any understanding of the issue. FoEI's Climate Change program had not yet been able to reach out to civil society in the country, and was thus unable to benefit from debate with and the participation of Chilean people. What happened? In November 2007,

> FoE Chile/CODEFF sent a representative to one of FoEI's Strategic Planning meetings in Swaziland, where they learned more about FoEI's program and thematic areas: they had a particular interest in finding out about the Climate Justice and Energy (CJE) Program and FoEI's Communications work. In 2008, CODEFF began to participate actively in the CJE program, and other FoEI strategy and skill share meetings. They also sent activists to UNFCCC climate change negotiations (in Croatia, Ghana and Poznan) and to key mobilization events. (FoEI 2010)

But with the Chilean activists attending numerous climate-centered meetings, and with the reframing as a climate justice campaign, it now enthusiastically hosts a 'strong national program'. FoE Chile's current position provides the following exemplar:

> We believe that the origin of climate change is in the unsustainable neo-liberal economic model that prevails in our country and much of the world. Solutions must recognize this and promote substantial changes in the patterns of socio-economic and human-nature relationships. Therefore in our country we promote the development of public policies on climate change. (translated from the Spanish FoE Chile 2010)

In this view of climate change, it becomes something far more than creating energy efficiencies and savings – by pursuing energy solutions which are not carbon-based – to an issue about key fault-lines in capitalism and its markets which have served the interests of the North and its intermediaries in the South for too long.

In the Asia-Pacific region, this climate justice frame has also taken root. WAHLI – FoE Indonesia – now has a climate campaign, whereas earlier in the millennium, it deliberately decided not to engage with such a campaign. Despite this move from rejection of climate change to an accommodation of its principles, WAHLI's frame still provides evidence of a deep suspicion of the dominant climate change discourse, continuing to see this narrative as just another, continuing form of suppression and dispossession of its people's development and livelihood by the affluent world. Its position is worth quoting in some detail here:

> Don't Trade off Our Climate ... Government of Indonesia makes this country as 'Carbon Toilet' for developed countries, through the mechanism of carbon-offset trading, and the addition of new debt.

> Development model of northern countries which is energy, land, and water greedy, and also exploits cheap labor, is actually the main cause of climate change catastrophe. Unfortunately the model is also adopted by developing countries such as Indonesia and believed to be the model of future development. Climate change has been diverted into a new legitimate tool to re-master the natural resources in developing countries as well as take control of the country's territory by developed countries. This was found in the scheme of mitigation in the forestry sector (Reducing Emissions from Deforestation and Degradation/ REDD). REDD offers the scheme to sell 26.6 million hectares of Indonesia's natural forests with trees, animals, plants, soil, water source, space of social interaction, and the entity of indigenous community in the area, only for 12 IDR per square meter ... False solutions offered in the climate change negotiations and implemented with debt support from developed countries, such as: REDD initiatives, carbon trading-offsetting mechanism, Clean Development Mechanism (CDM), transfer of dirty technology (agrofuel, nuclear, carbon capture storage) and genetically modified seed project on behalf of food security in drought, tornadoes and climate change. These false solutions will not reduce greenhouse gas emissions or save millions of small farmers, fisher folks and indigenous communities from the impact of climate change that have occurred at this time. The false solutions actually exacerbate the ongoing land conflicts, human rights violations and overlapping cross-cutting areas. (FoE Indonesia 2010)

By 2010, there is also ample evidence of South-North climate identity building, with Northern-groups accommodating a Southern, post-colonial voice into their predominantly post-industrial (and, with reference to non-humans, partly post-materialistic) campaign framings. The impacts of climate change on the Earth's poor are now trumpeted across position statements of European groups as a key determining factor informing climate action.

Thus, within FoE groups, climate change has moved from being seen exclusively as an issue of Northern groups, to a position which is not only shared by Southern groups but, in part at least, informed by Southern groups. This is not to argue that all differences in framing fade away. For example, although 'old-style economics' does now get a mention under the ubiquitous rubric of 'resisting neo-liberalism', for the Northern groups, the C word – Capitalism – still does not get a mention. In the case of the Latin American groups, however, capitalism is consistently named and ousted as the foremost climate villain. This does not imply that there are not groups within the UK, or other parts of Europe and North America, which directly critique and mobilize against capitalism (such as Climate Camp in the UK), but they operate outside of the organizational domain of FoE. Indeed the purpose of using the anti-neo-liberal catchphrase in campaigns – whether it is used by union, religious or green groups – was partly because of its ambiguous nature, as it allows the incorporation of a wide diversity of ideological frameworks. In most parts of the North, it is translated as 'anti-corporates' (changing and challenging corporate behavior, and demanding an increased role of the state and civil society), rather than anti-capitalism. Indeed, this became a key bone of contention at several general meetings. For example, at the 2010 Bi-annual General Meeting (BGM), Richard Navarro (representing FoE El Salvador) argued strongly that climate change could not be adequately addressed if capitalism itself remained unchallenged. Several of the Central and Southern American FoE groups commonly articulate their opposition to capitalism as the key climate villain. The work of Otros Mundos Chiapas (which became part of the FoE Confederation in 2008) is a good example. It is an organization that focuses on providing education, awareness and support about capitalism in Chiapas, the southernmost state of Mexico (FoEI 2010). Gustavo Castro, one of the founders of Otros Mundos Chiapas, talks about the relationship of climate change and other pressing environmental issues to capitalism:

> Mere awareness does not lead to immediate changes. Because of this, they are also committed to supporting, strengthening and creating movements and social processes to deal with the problems of capitalism within indigenous and poor communities. They have been involved in the creation of networks and campaigns against neoliberalism, dams, mines, against the foreign debt, militarization, against GMOs, monoculture plantations (mainly eucalyptus and palm oil) and against TNCs, and FTAs and Climate Change. (FoEI 2010)

Of course, as always, the North-South dichotomy is far from perfect. In the North, FoE Australia does sometimes use the 'capitalism' word, but places it within more anarchist ideological frames (Bello 2008). Amsterdam-based FoE International, usually unwilling to take a direct public position itself against capitalism, is also sometimes comfortable sitting alongside those groups which directly confront capitalism. At the people's summit on climate change in Cochabamba in Bolivia, a group of Friends of the Earth climate justice campaigners, led by Chair of FoE International, Nnimmo Bassey wrote this blog post:

> Evo Morales of Bolivia did not mince words yesterday when he diagnosed the root cause of climate change as being capitalism and all that it entails. President Morales stated that, in fact, Capitalism is the 'number one enemy of mankind.' He sees a sustainable future being possible only through 'actions of solidarity and complementarities as well as equity and the respect of human rights, right to water and biodiversity – the Rights of Mother Earth – a new system of rights that abolishes all forms of colonialism.' The President was speaking at the formal opening of the first-ever World Peoples Climate Change Summit (CMPCC). (Lee 2010)

Capitalism aside, the emergence of Nnimmo Bassey as President of FoE International (well known in the past for his work as Executive Director of Environmental Rights Action FoE Nigeria) provides us with another angle with which to trace the emergence, evolution and metamorphosis in the framing of the climate campaign across the global organization. Although the poorest countries in Africa still inhabit the non-climate change periphery, other African countries like South Africa and Nigeria are actually taking the lead on the core countries' climate campaigns. Nigeria is now a key player in both promoting its own agenda within FoE networks, as well as advancing – through Bassey's appointment – the international agenda throughout Africa and the rest of the confederation. In an interview with us in 2008, Bassey discussed the early problems with FoE International's climate agenda in Africa, but also mentioned that these issues have now been largely overcome:

> Before now there were a lot of limitations, but right now with the strategic planning process and the redefining of the programmes, the limitations have more or less been removed ... Before then, the way the programmes were formulated were not very accessible to us, and useful for our campaigns ... Some of the campaigns were very theoretical, in terms of [unclear, talking over each other 22:10] ... we campaign on that ... [However] The approach to some of the campaigns, for example climate change, was more about following international conventions and debates and negotiations. That doesn't really situate things on the ground. But now the programmes are reformulated, we find that it's really grass roots, and that's what we do. (Bassey: 2005)

In fact, the specific case of Nigeria is particularly salient here, as it demonstrates how, at first, the climate frame was not utilized at local or national levels (particularly in anti-mining campaigns against Shell). Instead, over time, the climate frame entered the fray, and has now become a dominant (if not *the* dominant) ideological scaffolding which FoE Nigeria now utilizes to describe their continued involvement in actions against Shell.

Repertoires for change – climate change and energy

In this chapter, we will use the bipartite ordering nomenclature of *insider* versus *outsider* repertoires. Under these broad banners, there are a range of different tactical and strategic tools worth mentioning: lobbying and legislative change; legal approaches; education programs; mass mobilizations; and service provision.

Insider politics: appeals-to-elites

As the *rule of law* is pivotal to defining liberal democratic political repertoires, legal tactics and strategies are numerous amongst FoE groups across the geopolitical board. FoE Argentina is very active, for example, pursuing legal actions, particularly at the domestic level and have 'initiated legal proceedings against those most responsible for global warming' (FoE Argentina 2010, translated from Spanish). FoE Nigeria have also used both domestic and international law as a tactical avenue in its fight against Shell, and now climate change. A more recent development is reminiscent of the Argentinean approach, with the construction of the concept of a *climate criminal.* In this interpretation, the rules of the current hegemonic game are accepted, but certain *criminals* are breaking the law – instead of there being systemic flaws in capitalist societies. As eluded to in the earlier poem by Nnimmo Bassey, climate criminals will be dealt with in a newly constructed 'climate tribunal'. In this classical liberal manner, most citizens are construed as 'good', and all that really needs to be done is to control and discipline the anti-social minority.

The legal approach of Friends of the Earth Australia in relation to climate has been most interesting. It has used foci on *environmental security* and *debt* to construct the notion of an *environmental* or *climate refugee. Climate debt* is a post-colonial environmental position – where industrialized countries have over-exploited their 'environmental space' in the past, having to borrow from developing countries in order to accumulate wealth, and accruing ecological debts as a result of this historic over-consumption. Friends of the Earth Australia maintains that climate refugees add to the 'climate debt' owed by the global North to the global South, due to the 'unsustainable extraction and consumption of fossil fuels'. Furthermore, because refugees are among the world's most vulnerable people, the protection of their rights is depicted as the principal concern in responses to climate change (Doyle and Chaturvedi 2011b).

Having established the existence of environmental refugees, FoE Australia has worked hard to have the United Nation international definition of 'refugee' (created in 1949) extended to include those who are forced to leave their homes due to climate change associated problems. Their climate justice campaign website reads (FoE Australia 2010a):

> For many years our climate justice campaign has forged the agenda on the human rights dimensions of climate change, most notably in the realm of climate refugees.
>
> Friends of the Earth has welcomed the call from Greens senator Sarah Hanson-Young for Australia to create a new visa category for Pacific Islanders affected by climate change ... Accepting climate refugees must be a central part of a meaningful response to climate change. (FoEA 2010b)

Friends of the Earth International (FoEI) increasingly endorses the Australian approach, placing the climate refugee within broader discussions on addressing climate debts by making reparations. In a report made by the organization, Davissen and Long make FoEI's critical position clear:

> The global North, as the major greenhouse polluter, bears a significant responsibility for this disruption. Accordingly, we believe that the North must make reparations. In practical terms, this will mean we must make room for environmental refugees, as well as changing policies that contribute to the creation of more refugees. (Davissen and Long 2003: 8)

Other liberal democratic, insider repertoires include appeals-to-elites campaigns (Martin 1993), which are common across all of the Confederation. More time and effort, however, is placed on these approaches in the North. Methods of indirect influence are the stock trade of movement organizations within liberal democracies. These include lobbying political parties, the state, international regimes for legislative change, and arguing for changes to the law at all levels though various judicial and quasi-judicial processes, and are the everyday grist for the mill for FoE climate activists. There is no better example of the traditional approach to lobbying than FoE USA, who largely confine their role to this textbook definition of the role of civil society in 'pluralist' democracies. In an interview with Elizabeth Bast, this focus on indirectly influencing domestic legislative politics in relation to climate change could not be clearer. It also accounts, in part, for the relative invisibility of FoE USA in the wider organization, and its lack of overall sustained contribution to the organization's work at an international level. Despite the USA's enormous power in global affairs, its civil society remains thwarted by navel-gazing and lacks an ability to challenge structures laid down for it by elites within its national sphere. FoE USA is just one example of this limited tradition (Doyle 2005). On FoE USA's relationship with local community groups, Bast states: 'It's more grass tops, I would say' *(as opposed to grass-roots)*. She adds that:

> An example from our work this year on climate change: there was a big climate change bill in Congress. We still haven't passed any legislation to reduce greenhouse gases, but there is now talk about it at the federal level. And so there was a big piece of legislation that was introduced and we thought it wasn't good enough, it wasn't going to get us where we needed to go. So we did an analysis of the bill and found that half of the money in the bill could potentially go to the coal industry, that was coming out of the auction revenue from reducing climate emissions. We put that out on blogs, in whatever media we could find to pick it up. We said we've got to be better than this, this is not going to solve the problem. Along with that we did other analysis on where the other parts of the revenue from the bill would go, and trying to push for revenue to go to better places, whether that's reducing energy costs for low-income families in the US, or international adaptation, or clean technology, or just investment in wind and solar and not into nuclear and bio-fuels and whatever. And so taking that analysis, pushing it out to media, but also going directly to Congressional offices and saying this is what's happening, we need to make this bill much stronger and much better, both in the way the revenue would be distributed, but also the bill wasn't strong enough in terms of the emissions reductions, we're saying this isn't actually going to solve the problem anyway. So we've got to do better. (Bast 2006)

Friends of the Earth Ireland has also been a predominant campaign body pushing for climate legislation at the national level. It also had cross-party support before the 2011 general election and is part of the new government's 'Programme for Government' (Friends of the Earth Ireland 2011). In this manner, FoE Ireland has had enormous access to state processes, and its policies became – for a short time before the dissolution of a coalition government (which included the Green Party) in 2011 – *part* of the state. This ready access to decision-making at an elite level may account, in part, for the virtual absence of the climate change discourse and repertoires at the public level. In fact, this separation of FoE Ireland from the grass-roots in Ireland regarding climate may even be starker, if that is possible, than the US account already provided.

In many ways, FoE Ireland enjoyed *insider* status with the state under the Coalition Government. Similarly, in the UK, FoE EWNI, at times worked closely with the Blair Labour Government, and operated under relatively 'friendly' conditions. The Big Ask campaign brought Friends of the Earth groups from 18 countries together, all with the same 'big ask': 'That their governments commit to reduce carbon emissions, year on year. Every year. In the UK the campaign has led to the groundbreaking "Climate Change Act". This has been followed by similar legislation in Scotland' (FoE EWNI 2010c). Obviously, all that needs to happen is for a change in government – as was the case in Britain in 2010 – for these relationships to change quickly and dramatically. This impacts immediately and directly on the selection of repertoires, reflecting the changing nature of the political opportunity structures.

Outsider politics: mobilization

Of course, in many more peripheral parts of the Confederation, insider politics is rarely an option. Instead of appealing to elites, many campaign repertoires are designed to *mobilize* opposition to the state and/or corporations. The distinction between appeal-to-elites and mobilization also works hand-in-hand with demarcations between the cosmopolitan and notions of solidarity. As discussed earlier, many of these decisions to select certain campaign repertoires over others emerge from the polity itself. There is no better example of this than the climate campaign. In both North and South, there is an emphasis on education. But as the 'community' is constructed so differently in either case, the very definition of what education entails is also fundamentally challenged.

In Chapter 2 of this book, we alluded to the fact that climate – viewed from a certain angle – is the ultimate western issue, because it is constructed from western science itself. In the North, FoE activists often see the answer to the problem as residing in their abilities to communicate this science (or a set of ethics in the case of religious organizations) to the public. The contention is that once the 'good science' or good ethics has been understood by the masses, then the problem can be solved. In Chapter 2, we referred to FoE France, as it lists its 'means of action', explaining this approach in terms of both food sovereignty and climate change: 'The organization of campaigns of sensitization (*campagnes de sensibilization*) to inform the citizens so that they change their practices' (FoE France 2010 translated from French). In this manner, the majority of people engage in '*wrong thinking*' (almost in the Marxist sense), and need to be re-educated to comprehend the '*true knowledge*' of human-induced climate change.

Education becomes the key issue of repertoire here – education as empowerment, versus education as 'right thinking'. In the South, education programs are designed to *empower* local communities. These communities already possess *true knowledge* but, through centuries of colonialism, these knowledge systems and voices have become muted, if not silent. All that is now needed, these activists argue, is to recognize these knowledge systems, and to assist in their vociferation. SAM/ Friends of the Earth Malaysia is engaged with this type of program:

> SAM/Friends of the Earth Malaysia has a long-standing collaboration with fisherfolk in Penang working with them to protect marine areas and ensure their livelihoods. The impacts of climate change on marine resources and increase in extreme weather events have led SAM to take practical action to support local Penang communities in mangrove restoration as well as undertake an education and awareness raising campaign on climate change causes and its projected impacts. (Nizam 2008)

Also, *education as mobilization* is often housed in terms of technical expertise, which is provided to communities to allow their innate knowledge systems of self-help to be invigorated. In this way, much climate education in the South is more about climate adaptation (after-the-fact) than climate mitigation (before-the-fact). FoE Brazil's climate campaign is a good example of this 'education as empowerment' approach.

For example, FoE Brazil has supported community projects linked to the National Movement of Struggle for Household and Shelter (MNLM) including:

- Horta Jardim Gordo, a collective urban garden managed by a local community association, which has been going for some time but needed reinvigorating. FoE Brazil gave technical assistance relating

to growing food, building a playhouse for children, and starting a rainwater-harvesting project. The garden is now producing food for 27 families, and provides activities and work for young people who are vulnerable to violence and drug dealing.
- At a 'house of passage' at Av Padre Cacique, community leaders support families who are resettling unused buildings, helping the families to generate income. FoE Brazil supported a three-day training programme, teaching people how to build rainwater-harvesting systems, grow medicinal herbs and make compost, as well as focusing on art and communication skills (FoE International 2010).

This approach to education in the South is also strongly evocative of the distinction made in the previous case studies between religious organizations in the North as lobbyists versus those in the South as *service providers*. Much of these education programs build the foundations upon which service provision can become a reality. Haiti Survive/ Friends of the Earth Haiti has been working on climate change research and impacts for a number of years, identifying *adaptation* actions to help communities cope with the effects of climate changes:

> A priority project that Haiti Survive is currently implementing is a community rainwater harvesting adaptation project to collect and store water for the dry season. This project is reinforcing food sovereignty activities in the local communities, increasing their resilience to impacts of climate change on food systems. (FoE International 2010)

Obviously, mobilization is not just about empowering communities through education, but may also take the form of simply a show of numbers, a show of force. Mass demonstrations are experienced in every country across the Confederation but, as mentioned in previous chapters, are a more typical response by activists engaged in *outsider* politics. In El Salvador, the Movement of Communities Affected by Floods (supported by Friends of the Earth El Salvador/CESTA) took part in a 100km 'walk for life' on 23 February 2009, demanding action for climate-affected peoples in El Salvador.

In many ways, this last case study illustrates the journey FoEI has taken, and the path it continues to tread. Within the Confederation, in the beginning, climate was seen as very much the exclusive domain of those FoE groups operating in the minority world – so much so, in fact, that FoE Equador listed this as one of the reasons it resigned from the organization. To its enormous credit, over time, FoEI has successfully managed to incorporate a very different post-colonial climate change understanding into its dominant international position. Geopolitical divisions in wealth and equality are now central to its mission.

Of course, FoE International is also a big player in choosing traditional appeals-to-elites methods such as lobbying – in all its forms – as a key campaign device. This is reflected in its attendance at numerous international diplomatic regimes, meetings, conferences or summits. At some of these climate-oriented meetings, FoE operates with almost *insider* status, seen as a 'responsible' NGO; whereas at others, it is seen as an *outsider*. In recent times, it has actually experienced the ignominy of being rejected from attending certain meetings. In fact, there is no better example than Friends of the Earth activists being blocked from attending the Copenhagen climate summit.

There is no doubt that at Copenhagen, FoEI's increasingly 'climate justice' position, based strongly on issues of geopolitical inequality (and making claims for a differentiated treatment of nations from the global South using the aforementioned rhetoric of 'climate debt' and 'climate reparations'), directly challenged the Northern climate hegemony of the western European nations. If FoEI continues to become more and more successful in genuinely representing Latin and South American, African, and Asia-Pacific interests (those from the lands of the Have-Nots), then it may have to increasingly expect an *outsider* status at international fora. FoEI may also be forced to pursue other forms of campaign repertoire, less designed to *indirectly* influence elite decision-makers. Ultimately, with outsider status, the powerful are beyond ear-shot.

Conclusion

Protest actions exist in both the North and South – whether they house themselves in Union movements, Church and religious groups, or environmental groups – but protest actions articulating climate change in the North usually mean a 'coming together' of voices of opposition on a particular day. In this manner, in the North, the *community of solidarity* is reconstituted at a moment in time, whilst in the South, communities protest where they already exist. They do not go home to a non-political space after the protest is over, they are protesting *in it.*

In the mostly emancipatory climate change campaigns we have written about here – whether in union, religious or environmental movements – these alternate constructions of what constitutes 'community' across the North-South divide, also impact upon our understanding of 'education' repertoires. In climate campaigns, both Northern and Southern activists are involved in mass education programs. In the North, the activists themselves believe that they have a source of elite, scientific knowledge or green ethics that must be imparted onto a largely 'ignorant' atomized populace. Once the population has been delivered the 'true message' of climate change (and their *capacities* have been adequately expanded into a global soul), then hopefully, people will pressure governments to adopt the political will to challenge carbon-reliant societies. In the South, with the 'community' already in existence, there is an innate respect for localized systems of knowledge. Education becomes a tool of empowerment for the people, rather than challenging the fundaments of their knowledge.

Global emancipatory movements have so far failed to identify relevant and compelling global targets for action and strategies for moving beyond the state. The meta-narrative of climate change – even as climate justice – may even be a step backwards, subsuming all human/environmental struggles under it, obscuring the material and power inequalities that North/South categories *are* so helpful at illuminating.

8
Making 'Climate Futures': Power, Knowledge and Technologies

Introduction

Our critical geopolitical analysis in the book has shown thus far that a number of 'climate futures' are competing with one another for greater salience, legitimacy and authority. Each seems bent upon proving its 'presence' by canvassing itself as more effective in countering 'global emergency', while claiming at the same time the high moral ground. Our key argument in this chapter is that discursively speaking, the idea of 'Climate Change', despite the overwhelming scientific evidence that it demands and deserves serious policy planning and effective action, is being slowly but surely turned into a site of shadow boxing where a variety of actors, institutions and agencies are implanting their own maps of meaning on spaces (terrestrial, oceanic, atmospheric) that they perceive as the most 'strategic', in pursuit of their respective geopolitical and geoeconomic agendas. Mike Hulme (2009: 340–341) is quite persuasive in his astute observation that it could be most revealing to

> examine climate change as an idea of the imagination rather than a problem to be solved. By approaching climate change as an idea to be mobilized to fulfill a variety of tasks, (the pursuit of profit, national security, human security, climate justice etc.) perhaps we can see what climate change can do for us rather what we seek to do, despairingly, for (or to) climate.

A more nuanced interpretation of this point will also direct our attention to the implications of the rhetoric and reality of climate change for the dominant understandings and framings of space, scale and power.

The British geographer Klaus Dodds (2012: 14) has argued that, 'Acting in advance of the future is an integral part of liberal-democratic life whether it is in the fields of climate change, terrorism, and/or trans-national epidemics.' It could be illuminating therefore to know how a particular future is made 'known and rendered actionable' in both temporal and spatial terms, and what consequences (and for whom) might follow 'from acting in the present on the basis of the future'. Equally important in his view is a critical examination of who and what are included in as well as excluded from various framings of a future:

> Making the future potentially actionable depends inter alia on a series of objects, practices and effects such as the generation of insights, trends, scenarios, and modeling; the production and circulation of images and reports; and the mobilization and distribution of anxieties, fears and hopes (Dodds 2012: 15–16).

In this chapter, by means of concluding this book, we critically examine three categories of climate futures, described by us as 'negotiated', 'engineered' and 'resisted'. The reason we have decided to focus more sharply on these three in this concluding chapter is because we feel that whereas the first two (and the others touched upon by us in earlier chapters) are representative of the most dominant logics behind the contemporary climate change discourses, namely that of state, neo-liberal market, science-technology, and military-security, the third represents the subaltern reasonings emanating largely (but not only) from the global South. All of them take climate earth science as their major evidence or *the* reference point in support of their narratives, while engaging in their own ways with the geopolitical issues of space, scale and power. Some of the questions that we try to address are: Who are the movers and shapers of a particular climate future and what kind of interests and agendas do they represent? Can there be major trade-offs between climate futures? What kind of alliances do we see emerging among various future narratives?

Negotiated future: climate diplomacy, common but differentiated responsibilities and respective capabilities (CBDRRC) and global governance

On the face of it, the argument that an internationally negotiated climate future could save 'humanity at risk' from unprecedented chaos and catastrophe through consensus based international climate diplomacy sounds quite persuasive and legitimate. The international system after all is still state-centric. Hopefully climate diplomacy might also fulfill the widely shared hope for what Daniel Innerarity (2013) would describe as a 'democratic management of risks' through a truly dialogic politics, but within a state-centric international geopolitical economy. Having said that, the challenges in the way of realizing this climate future, anchored firmly in the norm of CBDRRC through 'global governance', appear to be highly complex. First and foremost is the unsettled contestation over the nature of the phenomenon of climate change itself. Is climate change a 'global' or 'globalized' phenomenon? Let us reflect briefly on this point.

Dimitri D'Andrea (2013: 108) has shown how important the distinction could be between a 'globalized' and 'global' phenomena in the sense that, 'only the latter establish an *objective* foundation for the community of mankind. A globalized phenomenon causes an objective situation of interdependence (which can also be planetary); a global phenomenon leads to an objective situation of *community*.' His argument is that climate change is still a globalized and not global phenomenon and consequently 'it does not involve all the inhabitants of the planet or does not involve them all in the same way' (ibid.). In other words, a globalized phenomenon is one that:

> Involves human beings in general, but in ways that lead to or cause far-reaching differences between the interests of some and the interests of others, between those who risk damage and gain an advantage, on the one hand, and those who only risk damage on the other, between those who are exposed to radical damage and those simply exposed to a nuisance. In this sense, globalized space is *totally filled* (saturated) space, but it is not *homogenous*; a globalized phenomenon is a phenomenon to which no one is immune, but which for some is a resource and for others only damage, for some a choice and for others destiny, for some a catastrophe and for others a nuisance. (ibid.: 108)

Our key intention in this chapter is not to offer a detailed account of the origins and evolution of climate diplomacy within the framework of the UNFCCC, which has been recorded and analyzed in a rich body of scholarly literature. What we are chiefly interested in exploring, in the light of insights captured in the above quotation, is the manner (shifting patterns of political-spatial alliances and alignments) in which the challenge of globalized (but not yet global) climate change has been framed and debated within the UNFCCC. The UNFCCC came into force on 21 March 1994 with the aim of 'realizing stabilization of greenhouse concentrations in the atmosphere at a level that would prevent dangerous anthropogenic interference with the climate system'. It is further stated that, 'such a level should be achieved within a timeframe sufficient to allow ecosystems to adapt naturally [not through geoengineering] to climate change, to ensure that food production is not threatened, and to enable economic development to proceed in a sustainable manner.' Given the diverse nature and locations of its 195 member states, called 'Parties to the Convention', on the one hand, and considerable ambiguity around the issues of what constitutes 'dangerous anthropogenic interference' on the other, it is not in the least surprising that no specific time-frame was identified as 'sufficient' to enable ecosystems to adjust 'naturally' to climate change. 'The UNFCCC laid out the basic international, political, legal and normative architecture to address climate change. The core principle agreed under it noted that countries should protect the climate system on the basis of "equity" and in accordance with their 'common but differentiated responsibilities and respective capabilities' (Sengupta 2012: 102; United Nations 1992).

It was in Berlin, in 1995, that the First Conference of Parties (COP 1) was held and, as recalled by a senior diplomat serving as the head of the Indian delegation, right from the start,

> The negotiations reflected a deep North-South divide as well as major differences within both these groups. In general, developing countries pressed for an agreement based on equity, reflecting the fact that anthropogenic climate change was the result of cumulative emissions of greenhouse gases (GHGs) originating mainly in the developed countries. The developed countries, on the other hand, sought to minimize the link between commitments under the agreement and responsibility for causing climate change. (Dasgupta, 2012: 89)

The key concern of both the G-77 and China related to the implementation of current commitments. They argued that 'responsibility should not shift from Annex I to non-Annex I Parties' (ENB 1995: 3). The developing countries expressed the view that industrialized countries should commit themselves to more stringent emission reduction targets by adopting a legally binding protocol to the UNFCCC. This was met with strong resistance by a number of developed countries, particularly the USA (Sengupta 2012: 102). Against the backdrop of the release of the IPCC's Second Assessment Report 1995, which noted 'a discernible human influence' on the global climate, a decision called the 'Berlin Mandate' was finally adopted towards the end of COP 1. Supported by the EU, it called for negotiations aimed at a protocol with legally binding 'targets and timetables' to reduce emissions by the developed countries. India played a key role in taking this process forward, while insisting that the principle of equity demanded that, 'every human being had an equal right to the global atmospheric resources' and 'those responsible for environmental degradation should also be responsible for corrective measures' (Dasgupta 2012: 89).

However, one could find differences both within the global North and within the global South. A major divergence was particularly discernible among the OPEC countries led by Saudi Arabia. Anticipating highly adverse impacts of carbon mitigation measures on their petroleum exports, they were opposed to ambitious mitigation measures, even by the developed countries. On the other hand, the low lying island states, fearful of existential threat posed by the climate-induced sea-level rise, argued in support of the strongest possible climate regime. And as far as India, China and many other developing countries were concerned, they preferred a 'middle course' in an effort to hold together the 'G-77 and China group' (ibid.: 91–92).

What however turned out to be the hardest nut to crack was the following paragraph drafted by the Indian delegation, which after long and protracted negotiations was incorporated as Article 4, paragraph seven of the Convention.

> The extent to which developing parties will effectively implement their commitments under the Convention will depend upon the effective implementation by developed parties of their commitments under the Convention related to financial resources and transfer of technology and will take fully into account that economic and social development and poverty eradication are the first and overriding priorities of the developing country parties. (ibid.: 95)

By the time COP 2 met in Geneva from 8–19 July 1996, attended by more than 1500 from governments, intergovernmental organizations and NGOs participants (note the number would jump to about 10,000 in Kyoto for the next meeting), the US delegation for the first time appeared agreeable to the idea of supporting a legally binding agreement to fulfill the Berlin Mandate (ENB 1996: 1). The details of the nature and scope of the so-called commitment remained fuzzy however. The meeting concluded by 'noting the "Geneva Declaration," which endorses the IPCC conclusions and calls for legally binding objectives and significant reductions in greenhouse gas (GHG) emissions' (ibid.).

The Third Conference of the Parties (COP 3) to the United Nations Framework Convention on Climate Change (FCCC) was held from 1–11 December 1997 in Kyoto, Japan. The final outcome of one and half weeks of intense formal and information negotiations was the Kyoto Protocol, under which the Parties in Annex I of the FCCC agreed to commitments with a view to reducing their overall emissions of six greenhouse gases (GHGs) by at least 5 per cent below 1990 levels between 2008 and 2012. The protocol also established emissions trading, joint implementation between developed countries, and a 'clean development mechanism' to encourage joint emissions reduction projects between developed and developing countries.

The entangled nature of the interplay between geoeconomic hopes and geopolitical fears at this stage of climate negotiations, especially with regard to the notion of 'commitment', surfaced in several interventions made by China and India at Kyoto. China appeared convinced that, 'The developed countries are only interested in transfer of technical information, while developing countries deem technology transfer on non-commercial and preferential terms most important; and some countries emphasize market mechanisms.' 'India supported by Iran called for the operationalization of FCCC provisions relating to state-of-the-art environmentally sound technologies (EST), in the new legal instrument' (ENB 1997: 4). On 'voluntary commitments',

> India expressed its apprehension that the article would create a new category of Parties not established in the Convention. China said although the commitments were voluntary in name they would determine a level of limitation or reduction of anthropogenic emissions, imposing an obligation that did not apply to developing countries. The article endangers the non-Annex I status of Parties joining its activities and imposes new commitments on developing countries. (ENB 1997: 13)

The debate on voluntary commitments would flare up at again at COP 4 where India said, 'it was not implied in the principle of common but differentiated responsibilities' (ENB 1998: 3) and underlined what it perceived as a critically important distinction between luxury and survival emissions. The intellectual inspiration for the critically important distinction between 'luxury' and 'survival' emissions (reminiscent in some ways of the distinction Mahatma Gandhi had once made between the 'greed' and the 'need') came from a seminal article written in 1991 by Anil Agarwal and Sunita Narain of the New Delhi based Centre for Science and Environment (CSE), entitled 'Global Warming in an Unequal World: A Case of Environmental Colonialism' (see Dubash: 2012). Far from being in a denial mode on the climate change issue, Agarwal and Narain criticized the idea that 'developing countries like India and China must share the blame for heating up the earth and destabilizing its climate' and referred to a 1990 study published in the United States by the World Resources Institute in collaboration with the United Nations as 'an excellent example of environmental colonialism'. Both expressed the fear that censuring the developing countries for global warming would 'perpetuate the current global inequality in the use of earth's environment and its resources', and made a point which in our view is critically important for most of the Global South, especially in the context of neo-liberal globalization. Emphasizing the need for a 'strategy to improve land productivity and meet people's survival needs' they pointed out that 'development strategies will have to be ecosystem-specific and holistic'. Moreover, 'It will be necessary to plan for each component of the village ecosystem, and not just trees, from grasslands, forest lands and crop lands to water. To do this, the country will need much more than just glib words about people's participation' (ibid.).

To return to the Kyoto Protocol, it has been rightly pointed out that 'The politics of climate change – as demonstrated by the Kyoto Protocol process – raises dilemmas and paradoxes for politicians whose careers are framed by the demands of attending to a development model that must now come under scrutiny' (ENB 1997: 16). And what was indeed under close scrutiny in Kyoto was the same model of development that in the words of Walt Whitman Rostow 'begins with "take-off" and ends with a US-style mass consumption society'; a 'modernization' model that was marketed to the Third World of the 1950s and 1960s and bought by most of the decolonizing 'nation-states' feeling the burden of the revolution of rising social expectations. According to Kees Van Der Pijl (2014) 'Kennedy was impressed with Walt Rostow's *Stages of Growth* and in 1961 had him appointed deputy to his national security advisor, McGeorge Bundy. Later in the same year, Rostow became director of policy planning at the State Department.' In March 2001, the US, then the world's largest cumulative GHG emitter, announced its opposition to the Kyoto Protocol on the grounds that it believed it to be 'fatally flawed', as it would harm its economy and exempted major developing countries like China and India from similar emission restrictions (ENB 2001: 2; Sengupta 2012: 103).

The US decision to walk away from the Kyoto Protocol did galvanize in some ways a large number of the remaining convention parties, who then went on to adopt the Bonn Agreements at the resumed COP-6 (BIS) in June 2001 and later the Marrakesh Accords (Depledge 2005). Russia, until then a 'relatively minor player in climate politics', gained considerable geopolitical clout because the Protocol would not have entered into force had Russia refused to ratify. The Bonn Agreements came to the rescue of Russia, Canada, and Japan by freely allocating sinks to them under 'Land Use', 'Land Use Change' and Forestry (LULUCF) and thereby helping them to 'substantially reduce' their actual Kyoto targets (Afionis and Chatzopoulos 2009). On the one hand, one of the most outstanding things about the Protocol was that it aimed at setting up a 'market in carbon dumps as the main coordinating mechanism' (Lohmann 2005: 204), but at the same time this geoeconomic reasoning was heavily tempered with a geopolitical logic that territorialized the carbon cycle through the introduction of terrestrial carbon sinks (Stripple 2008).

The US decision with the Kyoto Protocol also created a geopolitical space for the EU to assert a leadership role in the UNFCCC process. Some analysts like Biermann sounded hopeful for Europe in the context of the global geopolitical and geoeconomic shifts in Asia's favor and argued:

> Europe should take this strategic position more seriously and should consciously strive to build up stronger, more stable relationships with the emerging great powers of Asia. In climate governance, Europe is forced to mediate its own interest in the climate issue with a multitude of non-European interests and negotiating positions, but also to forge a coalition of nations that is able to secure a credible, stable, flexible and inclusive governance system for the decades and centuries to come ... that Europe should take clear principled positions on a number of key issues; in particular the need to have a strong multilateral framework as the sole and core institutional setting for climate policy, and to accept the principle of equal per-capita emissions entitlements as the long-term normative bedrock of global climate governance. (Biermann 2005: 286)

Biermann (2005: 286) did acknowledge the possibility that the positions he was advocating to Europe might alienate the USA and at the same time make it more difficult for the US to rejoin the mainstream climate diplomacy. In his view, 'One [geoeconomic] way could be to link the European emissions trading system within the larger Kyoto context with actors and institutions in the USA' (ibid.). At this juncture in the evolution of the UNFCCC process (especially by the time of COP 15 in Copenhagen) a number of trends were quite visible. First and foremost, one could see serious disagreements 'between countries in terms of contributions to the stock of carbon in the atmosphere, industrial advancement and wealth, nature of emissions use and climate vulnerabilities' (Rajamani 2008: 1). Equally compelling, and overriding at the same time as this power-knowledge-power asymmetry, was the fact that despite the presence of several alliances and alignments formed on the basis of perceived convergence of interests, it was the major powers – both the present and the rising – that were aspiring to be *the* key movers and shapers of the agenda despite not being exactly on the same page. Some keen observers of the UNFCCC process, wondering about the painfully slow progress in the climate regime process would argue that what transpired at the 2009 Copenhagen Climate Change Conference provides enough evidence to suggest that, 'The slow progress of the climate change negotiations are due not just to the politics of the issue itself, but to the absence of a new political bargain on material power structures, normative beliefs, and the management of the order amongst the great powers' (Terhalle and Depledge 2013: 572). In their view,

> ...the governance-related processes of institutionalization and norm diffusion have failed to socialize rising powers (notably China) into the existing order, while at the same time failing to enmesh both 'indispensable' great powers (China and the US) into global governance structures. In doing so, they have neglected the impact of changes in the balance of power and of diverging normative world views. Instead of a more cosmopolitan order, a politico-military, and even economic, competition between China and the US has ensued, despite their interdependence. A process of order transition has emerged in which the material power structures and the normative beliefs underlying the Western order have been contested, with real repercussions for global governance regimes, notably on climate change. Owing to the broad great-power disagreement, global governance structures have become partly dysfunctional, and have encountered obstacles and deadlocks. (ibid.: 583)

Another take on this issue is that, with ecosystems (some of them highly stressed both on land and at sea) struggling to provide services to rapidly expanding global populations, the prospects of present and future resource scarcities have started galvanizing the fast growing economies, especially from Asia, in the direction of forming new alliances in climate change international negotiations (Hallding, Jurisoo, Carson and Atteridge 2013). Many were taken by surprise when Brazil, South Africa, India, and China decided to form the BASIC group (ibid.). The point worth noting is that, 'The coordination needed to align this heterogeneous group of countries cannot simply be understood in terms of a set of shared interests around climate policy' (ibid.: 608). What was unfolding was a major potential alliance among increasingly powerful countries. In other words,

> Although traditionally aligned with the G77 group of developing countries, recent strategizing as a group of emerging economies reflects their realization that there are insufficient global resources available to follow the same development pathway as industrialized countries. Hence, they must seek alternative growth pathways, which requires establishing common ground while also keeping track of each other's positions on important global issues like climate change. (ibid.)

It is important to note that the UNFCCC negotiations had by and large ignored various ecological concerns and issues of biodiversity, while focusing more on finance and technology, which, in any case were never mobilized in the way they could or should have been.

COP 15 could be seen as an outcome of contestations that had been brewing over a long time. The concise but controversial 'Copenhagen Accord' was negotiated in a hurry, on the final day of the conference, by a small but influential group of states lead by the United States, China, India, Brazil and South Africa (Sengupta 2012: 104). Unsurprisingly then, it invited a good deal of resentment (ENB 2009). Tuvalu, for example, found the so-called 'political agreement' lacking in terms of both a sound scientific basis and international insurance mechanism. Whereas 'Venezuela expressed indignation at the lack of respect for sovereign nations'. Bolivia, supported by Cuba, took offense at being given 60 minutes to decide an issue that could influence the livelihood security of millions of people. Costa Rica noted that in the absence of a consensus on the Accord, at most it could be issued as an information (INF) document. Nicaragua requested that 'the "Copenhagen Accord" be treated as a submission from those parties who negotiated it and issued as a miscellaneous (MISC) document; the COP and COP/MOP be suspended rather than concluded so that the AWGs' original mandates could continue; and a decision be taken to "mandate inclusive and transparent consultations, as appropriate" by the host country of the next session' (ENB 2009:7–8).

It was quite obvious that the Copenhagen Accord was dictated and driven more by considerations of geopolitical expediency than by the imperatives of legal norms. Consequently, considerable ambiguity surrounded the extent to which the so-called international community was willing to commit itself to legally binding emissions targets (McGoldrick, Williams and Rajamani 2010: 829). A serious and systematic pursuit of consensus based international climate cooperation with broader participation had been effectively stalled 'by the mutual conditionality of U.S. and developing country positions on climate change. The U.S. would not join an international climate regime unless and until major developing country emitters such as China and India were subjected to similar regulatory requirements. China and India would not join an international climate regime with binding commitments unless and until developed countries, notably the U.S., had shown willingness and ability to reduce their own greenhouse gas emissions first' (Skodvin and Andresen, 2009: 268).

It is useful to pause at this point and take note of the arrival of several international groupings – some new and some old – with diverse motives and agendas on the centre stage of global climate diplomacy during the passage from COP 1 to COP 16. They were the G-77/China, the Arab Group, the Environmental Integrity Group (EIG), the Alliance of Small Island States (AOSIS), the African Group, the Central American Integration System (SICA), the Bolivarian Alliance For The Peoples Of Our America (ALBA), the Least Developed Countries (LDCs), the Coalition of Rainforest Nations, the European Union (EU) and the Umbrella Group (see ENB COP 16). Despite the presence of a large number of NGOs and international organizations, the climate diplomacy remained (and remains still) in what John Agnew would call the 'territorial trap'.

At COP 17 in Durban in November 2011, one of the key outcomes included a decision by the Parties to adopt a 'universal legal agreement' on climate change as soon as possible, and no later than 2015. The contested manner in which the negotiations unfolded at Durban however revealed that there were a number of serious hurdles on the way including serious economic troubles in many countries, persisting political differences, conflicting priorities and divergent strategies for climate change mitigation and adaptation. Arriving at a consensus with regard to legally binding commitments for all without seriously compromising the principle of CBDR and equity turned out to be the hardest nut to crack.

> The first option deciding to develop a protocol under Convention Article 17 included elements on the content. The EU said that addressing the principle of CBDR 'in a contemporary and dynamic manner' is an essential component and suggested its inclusion. India, supported by China, suggested this option should be based on, and under, the UNFCCC and not involve reinterpretation or amendment of the Convention, with China suggesting that 'dynamic' interpretation of the principle may entail amendment. (ENB 2011: 23)

By the time COP 18 met in Doha in November 2012, one could also witness substantial disagreements between developing and developed countries over the complex issue of loss and damage. As pointed out by Hyvarinen (2013: 2),

> It is not clear if the UNFCCC negotiations on loss and damage have a realistic prospect of producing meaningful results. Constraints include concerns of developed countries about references to liability or compensation and the UNFCCC negotiating dynamics in general, which have not been conducive to progress for some time. Although it may not be politically feasible to pursue a different approach in the UNFCCC setting, vulnerable developing countries may need to consider the implications of acquiescing to inadequate rule development by participating in negotiations that give them little. This might even – in the wrong circumstances – carry the risk of loss of rights relating to claims for damage, if vulnerable countries do not exercise caution. For example, some small island states made specific declarations when joining the UNFCCC or Kyoto Protocol aimed at maintaining their rights under international law relating to state responsibility for the adverse effects of climate change.

The nascent debate on loss and damage shows how complex and difficult it might eventually turn to be calculating the 'costs' of loss and damage in post-colonial global South with long standing histories and geographies of colonial exploitation of nature and natural resources.

How will the discourse of 'loss and damage' work out vis-à-vis the principle of common but differentiated responsibility and respective capabilities?

Geopolitics of common but differentiated responsibility and respective capabilities

Ever since the inception of UNFCC, the principle of CBDRRC – with a rather intriguing geopolitical-normative interface – has remained both the most central and highly contested notion in the context of international climate diplomacy. According to the Preamble of the 1992 UNFCC, 'the global nature of climate change calls for the widest possible co-operation by all countries and their participation in an effective and appropriate international response, in accordance with their common but differentiated responsibilities and respective capabilities and their social and economic conditions.' On the face of it the principle 'captures the idea that it is the common responsibility of states to protect and restore the environment but that the levels and forms of states' individual responsibilities may be differentiated according to their own national circumstances' (Brunnee and Streck 2013: 589).

Geopolitically speaking however, and especially when approached from the perspective of the Global South, the principle of CBDRRC enables a non-western critical gaze to expose the hierarchical nature of the world order. The Global South's fear is that the Eurocentric rhetoric of 'sharing the finite global atmospheric commons' hides the harsh reality of the historical responsibility of the North with regard to overconsumption of this space, and denies the fact that

> on a per capita basis, the North has been consuming atmospheric space at a rate ten times greater than the South. And with the atmospheric commons being steadily and inequitably depleted, an equal per capita allocation of the diminishing remainder has become an increasingly inadequate proxy for equitable sharing of the global commons (Kartha 2011: 509).

The search for common ground anchored in the principle of CBDRRC, upon which we might build a new equitable, fair and just architecture of global climate governance has proved to be rather illusive. Whatever promises the rich and industrialized countries of the North have made over the past two decades, they remain largely undelivered. One of the key guiding principles so far of international climate diplomacy, carrying geopolitical implications and geoeconomic obligations, is shrouded in a great deal of legal ambiguity. As Lavanya Rajamani (2012: 125–126) succinctly points out,

> There are differences between Parties on the core content of this principle, including on the source, nature, contours, extent and relative weight of the 'common' responsibility shared by Parties; the source, nature, contours, criteria for, extent and relative weight of the differentiated responsibility individual or groups of Parties have; the significance and legal import of the term 'responsibilities'; and the significance, legal import, and relative weight of the term 'respective capabilities' as well as the relationship between 'responsibilities' and 'respective capabilities'. Flowing from these many considerations, there are differing views on the applications this principle lends itself to, the nature of obligations it entails, as well as the legal status and operational significance of the principle... but for now this principle, as also the climate change regime, is work in progress.

This worrisome persistence of mismatch between promise and delivery, against the backdrop of increasing volumes of greenhouse gases into the atmosphere, has led to a geopolitical reassessment of the 'success' and 'failures' in climate policy. A September 2013 a special issue of the journal *Climate Policy* on 'The Changing Geopolitics of Climate Change' has raised a number of pointers in this direction. According to the editorial (Streck and Terhalle 2013: 533),

> Inspired by Mackinder, this *Special Issue* suggests that climate change has introduced a new pivotal point in human development. Consequently, environmental governance and, more specifically, climate governance has become a matter of geopolitics in the 21st century. The increasing competition for resources (land, food, and fuel) by existing world powers has profoundly changed the context of international environmental governance.

The editors also highlight two key features that link the contributions to the special issue: 'First, there is an underpinning assumption that great powers will be the main actors in global environmental politics' and 'Second, the focus here is on the assessment and implication of approaches of rising and established great powers towards global climate norms' (Streck and Terhalle 2013: 534).

Having spent huge amounts of time, imagination and resources on discussions on mitigation and adaptation over the past two decades, the UNFCCC process has reached a point where the scope of climate diplomacy stands forcibly broadened to include and address the notion of 'loss and damage'. In a Guardian article about the COP 18 meeting in Doha, Ronald Jumeau, one of the negotiators from Seychelles, was reported to have said:

> If we had had more ambition [on emissions cuts from rich countries], we would not have to ask for so much [money] for adaptation. If there had been more money for adaptation [to climate change], we would not be looking for money for loss and damage. What's next? Loss of our islands? (Harvey 2012)

The article also reported that, 'The US had strongly opposed the initial 'loss and damage' proposals, which would have set up a new international institution to collect and disperse funds to vulnerable countries. US negotiators also made certain that neither the word 'compensation', nor any other term connoting legal liability, was used, to avoid opening the floodgates to litigation – instead, the money will be judged as aid' (ibid.).

The abovementioned conversation is representative of a 'clash of imaginations' grounded in both geopolitical and ethical considerations. While we agree by and large with what Simon Dalby has to say about the new geopolitics of the Anthropocene, we also visualize, at least in near to medium terms, a tension, bordering on conflict, between earth-centric view of climate security, emanating largely from the Global South, and atmosphere-centric views of climate security, being aggressively canvassed from the North. According to Dalby (2013: 40), geographers have taken a long time 'to recognize that the biosphere itself is now, in Latour's terms, effectively a hybrid of the artificial and the natural' and

> our new condition, this matter of life in the Anthropocene, is intimately interconnected with geopolitics, both in terms of how cultural representations of the earth have been generated, and in terms of how our technical practices have produced and enabled knowledge of the world in the process of trying to dominate and control it.

Yet it is not feasible (and perhaps undesirable as well) to downplay the harsh reality that for a vast majority of humanity, matters of life and death, literally speaking, are more intimately connected with the natural rather than the artificial (see Doyle and Risely 2008). This part of humanity is largely, but not exclusively, rooted in the earth and located in the Global South. A critical geopolitics of climate change needs to enlarge the scope of its 'relentless interrogation', in G O Tuathail's terms, to engage more seriously with the non-Western perspectives both *on* and *from* the Global South. As one of us, while questioning green grand narratives, has argued elsewhere (Doyle 2008: 315),

> ... the actual environmental issues on the ground are profoundly different in the South than the North. Movements, therefore, that surface in countries like India, Bangladesh, or Somalia – the majority world of the IOR [Indian Ocean Region], will be oriented around issues of basic environmental security ... the rights of people to gain access to the fundamental resources for survival: air, water, earth and fire.

On 16 February 2014, the US Secretary of State, John Kerry, addressed an audience in Jakarta (soon after visiting China on a similar 'climate mission') on the risks posed by climate change. It is obvious from the tone and tenor of his speech that while the US continues to have reservations about the principle of CBDRRC dictating and driving climate change diplomacy, it would like to garner the support of global South through a subtle geopolitics of emotions. Here are some excerpts from his long speech invoking both geoeconomic hopes and geopolitical fears in the same breath. Kerry's references to 'common', 'interconnections', 'responsibility', 'terrorism' and 'most fearsome weapon of mass destruction' deserve to be cited at some length:

> *Finally, if we truly want to prevent the worst consequences of climate change from happening, we do not have time to have a debate about whose responsibility this is. The answer is pretty simple: It's everyone's responsibility*. Now certainly some countries – and I will say this very clearly, some countries, including the United States, contribute more to the problem and therefore we have an obligation to contribute more to the solution. I agree with that. *But, ultimately, every nation on Earth has a responsibility to do its part if we have any hope of leaving our future generations the safe and healthy planet that they deserve ... Well, today in this interconnected world that we all live in, the fact is that hardship anywhere is actually felt by people everywhere. We all see it; we share it.*
>
> And when a massive storm destroys a village and yet another and then another in Southeast Asia; when crops that used to grow

> abundantly no longer turn a profit for farmers in South America; when entire communities are forced to relocate because of rising tides – that's happening – it's not just one country or even one region that feels the pain. In today's globalized economy, everyone feels it. *And when you think about it, that connection to climate change is really no different than how we confront other global threats.*
>
> *Think about terrorism.* We don't decide to have just one country beef up the airport security and the others relax their standards and let bags on board without inspection. No, that clearly wouldn't make us any safer. Or think about the proliferation of weapons of mass destruction. It doesn't keep us safe if the United States secures its nuclear arsenal, while other countries fail to prevent theirs from falling into the hands of terrorists. We all have to approach this challenge together, which is why all together we are focused on Iran and its nuclear program or focused on North Korea and its threat. The bottom line is this: it is the same thing with climate change. *And in a sense, climate change can now be considered another weapon of mass destruction, perhaps the world's most fearsome weapon of mass destruction.* (emphasis added)

John Kerry concluded his speech by assuring his audience in the global South that his country is ready to work together with Indonesia and the rest of the world, 'pulling in the same direction, we can meet this challenge, the greatest challenge of our generation, and *we can create the future that everybody dreams of* (US Department of State 2014; emphasis added).

As shuttle climate diplomacy continues, where do the climate negotiations stand now? According to some of those who have observed the process since the days of CoP 1,

> North and South have once again locked horns in the current climate change negotiations. In a nutshell, the developing countries are seeking *enhanced implementation* of the Framework Convention and the Kyoto Protocol, while the developed countries are pressing for a *new agreement* that would have the effect of amending or overwriting the basis provisions of the existing agreements. (Dasgupta 2012: 95–96)

For quite some time the focus on mitigation has been receding and the necessary wherewithal, especially financial resources, for adaptation is nowhere on the horizon. There is a new fear among some developing countries that 'some developed countries are now holding out threats of border levies or similar charges on exports of countries declining such coordination' (Dasgupta 2012: 96).

So far so good then for negotiated climate futures!

Geoengineered future: geographical politics behind 'climate engineering'

At a time when climate diplomacy seems to be heading towards new rounds of negotiations focused on highly complex and contested notions of 'loss and damage' – acknowledging rather grudgingly the failure so far in turning the promising rhetoric of climate mitigation and adaptation into reality –, the specter of 'climate engineering' (Hamilton 2013; Keith 2000; Morgan and Ricke 2010; Vaughan and Lenton 2011) has appeared on the horizon with an intriguing hope-fear interface. Bronislaw Szerszynski and Maialen Galarraga (2013: 2817) define geo-engineering

> as the intentional, large-scale manipulation of climate processes to offset the effects of anthropogenic climate change, brings together mundane technologies such as sulphate particles, mirrors, and olivine fragments with state-of-the art supercomputers and complex mathematical models, and the science of climate processes with engineering, studies of public perceptions, and the design of governance structures.

In their view geoengineering is better approached and understood as a 'complex heterogeneous assemblage' of diverse disciplines with varied intellectual premises and 'ways of knowing' (ibid.). The authors argue that issues related to the durability, affordability, and safety of geoengineering remain opaque at present and will have to be addressed through a multidisciplinary enquiry.

At the COP 12 meeting held in Nairobi in 2006 (against the backdrop of growing anxiety over putting into place the post-2012 climate mitigation regime), the issue of technology transfer turned out to be one of the most intensely debated issues. Developing countries (represented by South Africa, speaking on behalf of the G/77–China), emphatically argued that the key responsibility for mitigation rests with developed countries. The latter in turn underlined the need for all major emitters, including fast growing Asian economies, to take responsibility for limiting emissions (Okereke et al. 2007). Despite some movement forward, no landmark decisions were taken in Nairobi. The same year two major publications on geoengineering saw the light of the day. Paul Crutzen, Nobel Prize laureate, was one of the early advocates of undertaking a serious and systematic research on geoengineering especially in view of the fact that climate diplomacy was not moving fast enough in view of the lack of political will (Crutzen 2006). Tom Wigley (2006), the former director of Climatic Research Unit at the University of East Anglia, who later joined as a senior scientist at the National Center for Atmospheric Research in the United States, also argued that combining mitigation with geoengineering made good practical sense in view of the severity of anthropogenic climate change.

Making a major reference to the above mentioned essays by two eminent scientists, Alan Robock of the Department of Environmental Sciences at Rutgers University published a paper in the *Bulletin of the Atomic Scientists* offering '20 reasons why geoengineering may be a bad idea' (Robock 2008). According to Robock (2008: 15), 'These concerns address unknowns in climate system response; effects on human quality of life; and the political, ethical, and moral issues raised.' Robock begins his analysis by reminding his readers that the genesis of the idea of geoengineering can be traced back to the initial phase of the Cold War, when certain scientific communities in both the US and the Soviet Union invested considerable energy and finance funds into research aimed at controlling the weather. There were plans, for example, to use geoengineering for warming the Arctic (turning Siberia into a more habitable space), by damming the Strait of Gibraltar, and the Bering Strait. Once scientists became aware of rising concentrations of atmospheric carbon dioxide, the idea of artificially altering climate and weather patterns to reverse or mask the effects of global warming began gaining momentum (Robock 2008: 14).

On Robock's list of 20 reasons for concern, one finds concerns about unintended effects of unchecked emissions on regional climate systems and ocean acidification (because of a lessened focus on mitigation); ozone depletion (due to additional aerosols from geoengineering); and crops and natural vegetation and solar power (due to the lessened availability of sunlight). What about the environmental impacts of geoengineering? What about the margin of human error while dealing with such complex mechanical systems? What if there was an excessive climate cooling and an inability to retrieve aerosols from the atmosphere? Who will calculate the associated costs and who would decide the nature and scope of 'burden sharing'? Given that trillions of dollars are likely to be spent on geoengineering, Robock wonders: 'Wouldn't it be a safer and wiser investment for society to instead put that money in solar power, wind power, energy efficiency, and carbon sequestration?' (Robock 2008: 17).

In the context of geoengineering, the much-touted solar radiation management scheme (SRM) aims at counterbalancing heat-inducing rising concentrations of greenhouse gases by reflecting some of the inbound solar radiation back into space. A great deal of uncertainty and complexity surrounds the SRM. According to some analysts the political implications of SRM have not received the attention they deserve (see Szerszynski et al. 2013). They argue that,

> there is an urgent need to make explicit the particular way in which SRM is being constituted as a technology, to interrogate the embedded assumptions and sociopolitical implications of this constitution, to question whether it might encourage forms of politics that may be incompatible with democratic governance, and to explore the specific challenges that SRM might pose to democracy itself. (ibid.: 2810)

The highly skewed and uncertain manner in which SRM interventions through highly complex models, instruments and protocols might unfold are unlikely to result in a win-win situation for various stakeholders. The SRM is likely to raise highly complex issues of justice, redistribution, liability and accountability on the one hand, and 'it could generate a closed and restricted set of knowledge networks, highly dependent on top-down expertise and with little space for dissident science or alternative perspectives' (ibid.: 2812) on the other.

Non-western perspectives on geoengineering: resisting new dependencies?

The gaze of critical geopolitics on the issue of geoengineering need not stop here. It should also capture the non-Western perspectives on this issue. The IPCC Fifth Assessment Report (AR 5) has taken many analysts by surprise by including a section on geoengineering in the Executive Summary of Working Group I dealing with physical impacts of climate change. Having said that, there is no reason as of now to believe that the IPCC supports the idea of geoengineering. Its position is:

> Methods that aim to deliberately alter the climate system to counter climate change, termed geoengineering, have been proposed. Limited

> evidence precludes a comprehensive quantitative assessment of both Solar Radiation Management (SRM) and Carbon Dioxide Removal (CDR) and their impact on the climate system. CDR methods have biogeochemical and technological limitations to their potential on a global scale. There is insufficient knowledge to quantify how much CO2 emissions could be partially offset by CDR on a century timescale. Modelling indicates that SRM methods, if realizable, have the potential to substantially offset a global temperature rise, but they would also modify the global water cycle, and would not reduce ocean acidification. If SRM were terminated for any reason, there is high confidence that global surface temperatures would rise very rapidly to values consistent with the greenhouse gas forcing. CDR and SRM methods carry side effects and long-term consequences on a global scale.

The Bolivian Submission to the Joint Workshop of Experts on Geoengineering organized by the IPCC is quite illustrative of how many in the global South look at the future promise and perils of climate engineering.

> Geoengineering also has a history of belligerent use, and nothing guarantees that, in the current context, those that control such powerful technologies would not use them to hostile ends if they consider it convenient, although they may initially propose them as measures to face global warming. Therefore research on geoengineering can pose a possible future threat on fulfillment of the United Nations Environmental Modification Treaty (ENMOD), which prohibits the hostile use of environmental modification.

As pointed out by Simon Dalby (2013), there is yet another danger in geoengineering, besides the loss of transparency and oversight. With considerable ambiguity surrounding the question of who (military-security establishments or billionaire entrepreneurs or private-defense sector) is going to 'fix' the problem of soaring temperatures, the prospects of community-centric, community-driven initiatives to cope with weather-climate vagaries could be seriously compromised. It will be critically important to ensure that as highly elitist pursuits of geoengineering unfold they do not undermine the capacity of rural communities to pursue 'smaller scale ecological innovations' that are likely to 'buffer ecosystems against meteorological extremes while sinking carbon in ways that facilitate local livelihood security' (ibid.: 44).

And what about the military use of these technologies and the possibility of conflict with other treaties?

> The United States has a long history of trying to modify weather for military purposes, including inducing rain during the Vietnam War to swamp North Vietnamese supply lines and disrupt antiwar protests by Buddhist monks. Eighty-five countries, including the United States, have signed the U.N. Convention on the Prohibition of Military or Any Other Hostile Use of Environmental Modification Techniques (ENMOD), but could techniques developed to control global climate forever be limited to peaceful uses? (Robock 2008: 17)

The terms of ENMOD explicitly prohibit 'military or any other hostile use of environmental modification techniques having widespread, long-lasting or severe effects as the means of destruction, damage, or injury to any other State Party.' Any geoengineering scheme that adversely affects regional climate, for example, producing warming or drought, would therefore violate ENMOD. (Robock 2008: 17).

More recently, some defense analysts (Werrell and Femia 2014) have highlighted the role of the U.S. military in applying innovative technologies to realize climate secure future. The key argument here is that 'the United States has a long history of developing innovative technologies for improving war fighting that are eventually repurposed for civilian life. This includes society-altering technologies like the computer and the Global Positioning System (GPS)' (ibid.: 1). The most recent arrival on the list of such innovations is 3D printing. The non-traditional security threat to which, one is told, it could be usefully applied is disaster relief, search and rescue and various contingencies arising out of climate related events and natural disasters. It is further pointed out that, 'As with technological innovation, the military is already leading the way in recognizing and preparing for the security risks associated with climate change ... *The main preoccupation is that climate change may act as an 'accelerant of instability,' exacerbating others drivers of unrest such as water, food and energy insecurity*' (ibid.; emphasis added).

Resisting artificial climate futures: subaltern perspectives on climate present

Climate change scenario building from different vantage points, invoking hope on certain occasions and fear on others, will continue unabated and this is not in the least surprising. With state-centric climate change diplomacy not getting anywhere closer to a 'global governance' that would deliver legally binding solutions to a 'planetary emergency' based on equity and social justice, a steadily growing number of academics, policy makers and media experts seem inclined towards market-military based solutions to global warming that would be assisted by geoengineering. John Agnew's insistence that 'there is no such thing as a view from nowhere' stands fully vindicated by the ways in which the affluent and the powerful continue to deploy fear-inducing metaphors and narratives that illuminate the *consequences* of climate change and push the *causes* under the carpet. It is asserted by some, and for good reasons perhaps, that climate change is the issue of 'production'. That climate change is also (and perhaps more so) about the issue of 'consumption' is relatively underplayed. An overwhelmingly large number of scenarios, models and reports – despite being well meaning – either make a passing reference to earth-centric, day-to-day struggles of millions on the socio-economic and political margins of the emerging, globalizing capitalist system, or pay a lip service to them or make them completely voiceless and faceless.

As pointed out by us earlier in both this concluding chapter, and this book, for millions belonging to the 'present generation' in the 'Majority World' it is the still the 'Earth' that matters more than the 'Atmosphere'. Here are just a few facts to ponder over. According to the WHO, in 2012, 7 million people lost their lives due to polluted air – in most cases a by-product of unsustainable policies in sectors such as transport, energy, waste management and industry (Climate Action 2014). A vast majority of these casualties happened in Southeast Asia.

Conversely, there is no dearth of examples showing how communities on the socio-economic margins of the islands of affluence are resisting the assault of neo-liberal globalization on their land and resources (Doyle and McEachern 1997). D L Sheth (2004: 56) would describe these micro movements as a symbol of participatory democracy: 'a parallel politics of social action, creating and maintaining new spaces for decision-making (i.e. for self-governance) by people on matters affecting their lives directly.' Variously described as grass roots movements, social movements, non-party political formations, social-action groups and movement-groups, what these assertions from below share in common is a firm opposition against the post-political push of neo-liberal globalization. They demand more and not less politics.

Pankaj Gupta (2008) in his thoughtfully entitled paper, 'From Chipko to Climate' takes the village of Jardhar in Garhwal region (in the India state of Uttarakhand) as a case study to argue and illustrate that, 'the global conservation ethic and global development are, in certain circumstances, detrimental to local interests: they transfer costs from power urban centers and demand sacrifices from fragile mountain communities' (Ibid. 4). He refers to *Beej Bachao Andolan* (Save the Seeds Movement) which continues to resist the government sponsored schemes to spread the Green Revolution in the mountains in order to 'increase food availability in cities and to keep food prices low' (Ibid. 6).

The net result is that the community that once had over 80 varieties of rice and over 200 kinds of beans stands split as some local farmers, lured by the prospects of higher and faster yields, opted for hybrid seed varieties of rice. Abandoning subsistence ecology in favor of commodity production has wideranging implications for agricultural sustainability based on seed preservation, compost and crop diversity (Ibid). Gupta quotes (Ibid. 6) Vijay Jardhari (a leading voice of the *Beej Bachao Andolan*): 'a farmer's independence can only be ensured if he keeps his own seed, otherwise he is just a slave of the company or the government. What kind of new seeds are these that cannot be kept for the next crop?' In his view, 'In the 21st century, as climate change takes center stage in the global environment debate, forests—in order to fulfill their 'carbon sink' function—could be made even more inaccessible. Again, it is local communities like Jardhar that will be the vanguard of a revolution not of their making.' We would like to expand the scope of Gupta's argument by saying that along with the forests, the water bodies (the seas, rivers, estuaries, mangrove forests etc.), in order to fulfill their 'blue carbon sink' as well as 'carbon trading' functions, also might become increasingly inaccessible to millions in global South who depend on them for their livelihoods.

Final conclusion

The construction, then, of a *universal* we or a *global us* is detrimental to the welfare of the many, whether it is green, black, red or blue. It is tempting, at first glance, to argue that the post-political masquerade – which is liberal cosmopolitanism – is soft politics. Within this world-view which promotes the concept of one global citizenry, it is imagined that boundaries between the haves and the have-nots, the territories delineating the rich from the poor, melt away into the air. At its best, we can understand this world-view as *a* political, as spineless, as incapable of delivering welfare promises to anyone other than those already in positions of power.

Upon subsequent glances, the conclusions are not so generous. Of course, lines of demarcation never fade away completely. They exist always as a palimpsest of contested political territories: new maps of power are inscribed over old maps. New borders are being drawn which *re-territorialize* the planet. In this new green map, the lines between haves and have-nots do not disappear at all. Rather, denying the gross disparities in welfare between the affluent North and the global South (through the pursuit of apparently global post-political agendas), further strips power from the powerless, and strangles the voices of the poor. Issues such as Northern-inspired and constructed climate change; the rights of the non-human; and the welfare of future-generations are post-industrialist and/or post-materialist. They have little relevance to the everyday welfare of those inhabiting the global economy's peripheries. But worse, it is not just that they are irrelevant to the real issues of welfare and survival; by their position of primacy on global green agendas, far more fundamental issues such as food sovereignty, water security, and fair access to energy resources are lost in the liberal cosmopolitan noise.

And even less charitably, as so often is the case with liberal discourses, they politely mask more realist, ultimately conservative political agendas. For example, this 'depoliticized' space which imagines no rooted communities of space and place, fits in very easily with the neo-liberal economics of the radical libertarians. The global green citizen, hiding behind narratives of deliberative and representative democracy, is really none other than the global green shareholder. All green welfare issues are now imagined to be resolved using market-mechanisms, its decisions to be forged by green plutocrats. The state has a scarce role to play. The case of climate change discussed above is a stellar example of the way global purchasers, providers and consumers are going to *trade* their way out of global carbon-induced Armageddon.

Also on the right, of course, are the green conservatives. One can hear the increasingly raucous cries of the neo-Hobbesians and Neo-Malthusians, grabbing the aforementioned discourses of sustainable development and ecological modernization (Chapter 2), and refashioning them within the new language of *resilience*. This current 'green state' welfare policy push to create *resilient communities* is conservative at its core. It invents a green sphere which is premised upon the notion that we need to return society back to a steady-state, *before* the great ecological disruption. Indeed, the very definition of resilience is based upon the abilities of ecological communities to restore natural order as quickly as possible after disturbance. In the case of climate change, this means a policy switch to climate adaption policies, rather than mitigation. But this faith in a *green natural order* beyond humans is misplaced. It is an order which has never served the welfare interests of the majority of people on this planet, and never will (Catney and Doyle 2012).

Earlier in the previous chapter, we recalled a meeting of Friends of the Earth International in Abuja, Nigeria. At this meeting, when a European activist insisted that the Ogoni's enemy was climate change, rather than the specific dispossession of the people by Shell and the state, the Ogoni elder was generous in his response: he would accept this new climate framing so that it may help 'sell' his peoples' story in the more affluent world. From a position of powerlessness, we can understand the Ogoni Elder's willingness to 'reframe' his reality in a bid to get more support from world environmental NGOs. Probably the best thing about the local-level oil pipeline issues in Nigeria becoming transnational has nothing to do with the scientific substance of climate change, whether it is real or not (or whether our dependence on carbon can be reduced). But rather, this new transnational frame has helped bypass national governments, building a bridge between disenfranchised, peripheral local communities with a more affable (though more directly depoliticized and disconnected) transnational citizenry. The language of climate change, though largely politically sanitized and devoid of meaning in post-colonial societies, expedited this process. So the *local* speaks to the *transnational* when the *national* ignores them or, at worst, threatens coercion or enacts violence upon them. In the case of the Ogoni people, and other minority groupings in the Niger Delta who lived under the jackboot of a military regime, perhaps one of the ways for local people to survive (and derive hope for the future) is to bypass the nation-state, and reach out into transnational spheres.

But there may also be a loss when what is effectively a post-colonialist story of abject poverty is replaced with a post-materialist, post-political and post-industrialist one of the affluent world. Perhaps there is a human essence which disappears here, diluted and re-colored to a point and a time when the essence can no longer be identified. The stories of daily survival and dispossession are lost, replaced with more benign archetypal myths of distant floods and fires. Of course, for the majority world, it is hard to get excited about potentially rising sea levels measured by elite scientists in the North.

Armageddon came many years ago for the globally peripheral peoples of the global South, when their lands were first invaded and colonized by their European oppressors. The flood came long ago – it was their peoples' blood.

Bibliography

A

Ackerman, J. T. 2008. 'Climate Change, National Security and the Quadrennial Defense Review: Avoiding the Perfect Storm', *Strategic Studies Quarterly* (Spring): 56–96.

Adams, V., Murphy, M. and Clarke, A. 2009. 'Anticipation: Technoscience, Life, Affect, Temporality', *Subjectivity*, 28: 246–265.

Adler, J. 2007. The IPCC's Fragile Consensus, http://volokh.com/posts/1170434603.shtml, accessed 14 July 2013.

Afifi, T. and Warner, K. 2008. *The Impact of Environmental Degradation on Migration Flows Across Countries*, Working Paper No.5, UNU-EHS Working Paper Series, United Nations University, Institute for Environment and Human Security.

Afionis, S. and Chatzopoulos, I. 2009. 'Russia's Role in UNFCCC Negotiations Since the Exit of the United States in 2001', *International Environmental Agreements: Politics, Law and Economics*, 10(1): 45–63.

Agnew, J. 1994. 'The Territorial Trap: The Geographical Assumptions of International Relations Theory', *Review of International Political Economy*, 1(1): 53–80.

Agnew, J. 1998. *Geopolitics: Re-visioning World Politics*, London: Routledge.

Agnew, J. 2010. 'Emerging China and Critical Geopolitics: Between World Politics and Chinese Particularity', *Eurasian Geography and Economics*, 51(5): 569–582.

Ananthapadmanabhan, G., Srinivas, K. and Gopal, V. 2007. Hiding Behind The Poor, 12 November, http://greenpeace.org/india, accessed 2 August 2014.

Andrea, D. 2013. 'Global Warming as a Globalized Risk and Global Threat for Future Generations', in Daniel Innerarity and Javier Solana (eds) *Humanity at Risk: The Need for Global Governance*, New York: Bloomsbury: 107–119.

Argos, C., Reeves, H., Jouzel, J. 2010. *Climate Refugees*, Cambridge: MIT Press.

Ashton, J. and Burke, T. 2004. The Geopolitics of Climate Change, SWP Comments, http://www.swp-berlin.org/fileadmin/contents/products/comments/comments2004_05_ashton_burke_ks.pdf, accessed 31 August 2014.

Atkinson, J. and Scurrah, M. 2009. *Globalizing Social Justice: The Role of Non-Government Organizations in Bringing About Social Change*, Basingstoke: Palgrave Macmillan.

Attfield, R. 2003. *Environmental Ethics*. Cambridge: Polity.

B

Bailey, I. 2007. 'Neoliberalism, Climate Governance and the Scalar Politics of EU Emission Trading', *Area* 39(4): 431–442.

Bakshi, R. 2009. *Bazaars, Conversations and Freedom: For a Market Culture Beyond Greed and Fear*, New Delhi: Penguin.

Baldwin, D. A. 1995. 'Security Studies and the end of the Cold War', *Journal of World Politics*, 489(1): 117–141.
Balzacq, T. 2005. 'The Three Faces of Securitization: Political Agency, Audience and Context', *European Journal of International Relations,* 11: 171–201.
Banerjee, P. and Samaddar, R. 2006. *Migration and Circles of Insecurity*, Calcutta: Calcutta Research Group.
Barker, A. 2008. Climate Change Migrants: A Case Study Analysis Internship Project Report, http://www.adb.org/Documents/Climate-Change/Migration-Final-Report.pdf, accessed 23 December 2013.
Barnett, J. 2001. Security and Climate Change, Tyndall Centre Working Paper, http://www.tyndall.ac.uk/publications/working_papers/wp7.pdf, accessed 23 September 2013.
Barnett, J. 2003. 'Security and Climate Change', *Global Environmental Change,* 13(1): 7–17.
Barnett, J. 2007. 'The Geopolitics of Climate Change', *Geography Compass*, 1(6): 1361–1375.
Barnett, J. and Webber, M. 2010. 'Accommodating Migration to Promote Adaptation to Climate Change', Policy Research Background Paper to the 2010 World Development Report, Working Paper 5270, World Bank.
Barrero, R. Z. 2013. 'Borders in Motion: Concept and Policy Nexus', *Refugee Survey Quarterly*: 1–23.
Barry, J. and Paterson, M. 2004. 'Globalisation, Ecological Modernisation and New Labour', *Political Studies*, 52(4): 767–784.
Bast, Elizabeth [interview]. 2006, 'On relationship with local community groups', Friends of the Earth USA, by Tim Doyle.
Basu, S. P. (ed.) 2009. *The Fleeing People of South Asia: Selections from Refugee Watch*, London: New York: Anthem Press.
Bayer, P., Athanasiou, T. and Kartha, S. 2008. The Right to Development in a Climate Constrained World: The Green House Development Rights Framework, http://www.ecoequity.org/GDRS, accessed 31 August 2014.
BBC. 2008. 'Summit ends with climate vision', http://news.bbc.co.uk/2/hi/asia pacific/7497032.stm, accessed 15 December 2014.
Beck, U. 1992. *Risk Society, Towards a New Modernity*, London: Sage.
Beck, U. 1999. *World Risk Society*. Cambridge: Polity Press.
Beck, U. 2009. *World at Risk,* Cambridge: Polity Press.
Beck, U. 2010. 'Climate for Change, or How to Create a Green Modernity', *Theory, Culture & Society*, 27: 263–201.
Beck, U. 2013. 'Living in and Coping With a World Risk Society', in Daniel Innerarity and Javier Solana (eds) *Humanity at Risk: The Need for Global Governance*, New York: Bloomsbury: 11–17.
Beckman, W. and Pasek, J. 2001. *Justice, Posterity and the Environment*, Oxford: Oxford University Press.
Bell, D. R. 2004. 'Environmental Refugees: What Rights? Which Duties?' *Res Publica,* 10(2): 135–152.
Bello, W. 2008. 'Will Capitalism Survive Climate Change?', Chain Reaction (104) www.foe.org.au/resources/chain-reaction, accessed 22 November 2008.
Bidwai, P. 2003. South Asia: Immigrant Issue Sours Indo-Bangladesh Relations, IPS-Inter Press Service, http://www.encyclopedia.com/doc/1G1-19353244.html, accessed 23 December 2013.

Bidwai, P. 2009. 'The Climate Impasse', *Frontline* 26(17), http://www.frontline.in/static/html/fl2617/stories/20090828261710900.htm, accessed 1 August 2014.

Bidwai, P. 2012. *The Politics of Climate Change: Mortgaging Our Future*, Hyderabad: Orient Black Swan.

Biermann, F. 2005. 'Between the USA and the South: Strategic Choices for European Climate Policy', *Climate Policy*, 5(3): 273–290.

Biersteker, T. and Weber, C. 1996. *State Sovereignty as a Social Construct*, Cambridge: Cambridge University Press.

Birrell, A. 2000. *Chinese Myths*, Austin: University of Texas Press.

Black, R. 2007. Stark Picture of a Warming World, http://news.bbc.co.uk/2/hi/science/nature/6524325.stm, accessed on 13 October 2013.

Blazey, P. and Goving, P. 2007. 'Financial adaptation challenges for the insurance industry due to climate change', *Macquarie Journal of Business Law*, 4: 15–48.

Boano, C. 2008. FMO Research Guide on Climate Change and Displacement, Forced Migration Online (FMO) September, http://www.forcedmigration.org/guides/fmo046, accessed 31 August 2014.

Bodenham, P. 2005. 'Lifestyle: Operation Noah – The Community Climate Change Campaign', *Journal for the Study of Religion, Nature and Culture*, 10(1): 109–115.

Bolivian Submission to Joint Workshop of Experts on Geoengineering. https://unfccc.int/files/meetings/ad_hoc_working_groups/lca/application/pdf/bolivian_submission_on_geoingeneering.pdf, accessed 3 August 2014.

Bourbeau, P. 2006. Migration and Security: Securitization Theory and its Refinement, http://www.allacademic.com//meta/p_mla_apa_research_citation/0/9/8/1/3/pages98136/p98136-1.php, accessed 2 January 2014.

Bowen, M. 2007. *Censoring Science: Inside the Political Attack on Dr. James Hansen and the Truth of Global Warming*, New York: Penguin Group.

Boyle, J. 2012. A Mirage in the Deserts of Doha? Assessing the outcomes of COP 18. International Institute for Sustainable Development. http://www.iisd.org/pdf/2012/com_mirage_desert_doha_cop18.pdf, accessed 23 August 2014.

Brindal, E. 2008. Climate Change Refugees: The Forgotten People, http://www.foe.org.au/climate-justice/media/news-items/archive/front-page-news-feed-1/climate-change-refugees-the-forgotten-people/, accessed 3 December 2013.

Brito, R. 2009. 'Securitizing Climate Change: Process and Implications', ISA – ABRI Joint International Meeting Rio de Janeiro, Brazil.

Brown, O. 2007. 'Climate Change and Forced Migration: Observations, Projections and Implications', a background paper for the 2007 Human Development Report, *Fighting climate change: Human solidarity in a divided world*, https://www.iisd.org/pdf/2008/climate_forced_migration.pdf, accessed 12 December 2014.

Brown, O. 2008a. 'The Numbers Game', *Forced Migration Review*, (31): 8–9.

Brown, O. 2008b: Migration and Climate Change, IOM Migration Research Series Paper No. 31, 2008, Geneva: International Organization for Migration, http://www.iisd.org/pdf/2008/migration_climate.pdf, accessed 20 December 2014.

Brown, L. R. 2009. *Plan B 4.0: Mobilization to Save Civilization*, New York and London: Earth Policy Institute.

Brunnee, J. and Streck, C. 2013. 'The UNFCCC as a Negotiating Forum: Towards Common but More Differentiated Responsibilities', *Climate Policy*, 13(5): 589–607.

Brzoska, M. 2008. The Securitization of Climate Change and the Power of Conceptions of Security, http://www.allacademic.com//meta/p_mla_apa_research_citation/2/5/3/8/8/pages253887/p253887-1.php, accessed 20 July 2014.

Bull, H. 1977. *The Anarchical Society: A Study of Order in World Politics*, New York: Columbia University Press.
Burke, T., and Ashton, J. 2004. 'The geopolitics of climate change.' Unpublished paper presented by the authors at an SWP roundtable on climate change and foreign policy on 4 February 2004, http://www.envirosecurity.org/conference/background/ClimateChangeGeopolitics.pdf, accessed 29 October 2014.
Burleson, E. 2010. 'Climate Change Displacement to Refuge', *Journal of Environmental Law and Litigation*, 25(1):19–36.
Burroughs, W. J. 2005. *Climate Change in Pre-History: The End of the Reign of Chos*, Cambridge: Cambridge University Press.
Burson, B. 2010. (ed.) *Climate Change and Migration: South Pacific Perspectives*, Institute of Policy Studies, School of Government, Victoria University of Wellington.
Bury, J. B. 1960. *The Idea of Progress*, New York: Dover Publications.
Busby, W. 2007. Climate Change and National Security: An Agenda for Action, CSN No. 32, Council on Foreign Relations, http://www.cfr.org/publication/18034/academic_module.html, accessed 23 November 2013.
Buzan, B. 2012. 'Asia: a geopolitical reconfiguration', *Politique Etrangère* 77(2): 331–343.
Buzan, B. Wæver, O. and Wilde, J. D. 1998. *Security: A New Framework for Analysis*, USA: Lynne Rienner Publisher.

C

Cahill, D. 2008. 'Labo(u)r, the Boom and the Prospects for an Alternative to Neo-Liberalism', *Journal of Australian Political Economy*, 61: 321–335.
Cambanis, T. 2012. 'The amazing expanding Pentagon' The Boston Globe 27 May. http://thanassiscambanis.com/2012/05/25/the-amazing-expanding-pentagon/, accessed 24 November 2014.
Campbell, K. M. et. al. 2007. *The Age of Consequences: The Foreign Policy and National Security Implications of Global Climate Change,* Washington, D.C: Center for Strategic and International Studies.
Caritas Internationalis. 2009. *Climate Justice: Seeking a Global Ethic*, Rome: Caritas Internationalis General Secretariat.
Carrington, D. 2010. WikiLeaks Cables Reveal how US Manipulated Climate Accord, The Guardian, 3 December, http://www.theguardian.com/environment/2010/dec/03/wikileaks-us-manipulated-climate-accord, accessed 31 August 2014.
Castles, I. 2002. Letters to Dr Pachauri, http://www.economist.com/media/text/efhpdoc1.pdf, accessed 2 November 2013.
Castles, S. 1993. *The Age of Migration,* New York: Palgrave Macmillan.
Castree, N. 2003. 'The Geopolitics of Nature', in John Agnew and Katharyne Mitchell (eds) *A Companion to Political Geography*, Oxford Blackwell.
Catchlove, J. 2006. *Seeking Sustainable Solutions to Climate Change*, Friends of the Earth Adelaide, Adelaide.
Catney, P. and Doyle, T. 2011a 'The Welfare of Now and the Green (Post) Politics of the Future', *Critical Social Policy*, 31(2): 174–193.
Catney, P. and Doyle T. 2011b. 'Challenges to the State' in T. Fitzpatrick (ed.) *Understanding the Environment and Social Policy*, Bristol: The Policy Press.
Challen, S. 2010. *Migration in the 21st Century: How Will Globalization and Climate Change Affect Human Migration and Settlement? (Investigating Human*

Migration & Settlement), London: Crabtree Pub Co.Chaos, Cambridge: Cambridge University Press.

Charveriat, C. 2000. Natural Disasters in Latin America and the Caribbean: An Overview of Risk, Inter-American Development Bank Working Paper, no. 434:58, http://www.iadb.org/IDBDocs.cfm?docnum=788256, accessed 23 July 2013.

Chaturvedi, S. 1998. 'Common security? Geopolitics, development, South Asia and the Indian Ocean', *Third World Quarterly*, 19(4): 701–724.

Chaturvedi, S. 2012a. 'De(securitizing) the Ice: Circumpolar Arctic in "Global" Climate Change', in Pami Aalto, Vilho Harle and Sami Moisio (eds) *Global and Regional Problems: Towards an Interdisciplinary Study*, Aldershot: Ashgate.

Chaturvedi, S. 2012b. 'The Antarctic "Climate Security" Dilemma and the Future of Antarctic Governance', in Alan D. Hemmings, Donald R. Rothwell and Karen N. Scott (eds) *Antarctic Security in Twenty First Century*, London: Routledge.

Chaturvedi, S. and Doyle, T. 2010. 'Geopolitics of Climate Change and Australia's "Reengagement" with Asia: Discourses of Fear and Cartographic Anxieties', *Australian Journal of Political Science*, 45(1): 95–115.

Chaturvedi, S. and Painter, J. 2007. 'Whose World, Whose Order? Spatiality, Geopolitics, and the Limits of World Order Concept', *Cooperation and Conflict*, 42(4): 375–395.

Chen, X. 2012. *Social Protest and Contentious Authoritarianism in China*, Cambridge: Cambridge University Press.

Chesters, G. and Welsh, I. 2006. *Complexity in Social Movements: Multitudes at the Edge of Chaos*, London: Routledge.

Chimni, B. 2000. *International Refugee Law: A Reader*. New Delhi: Sage.

Chin, J. 2008. 'Coping with Chaos: The National and International Security Aspects of Global Climate Change', *The Journal of International and Policy Solutions*, 9: 15–16.

Chowdhury, A. 2009. The Coming Crisis: From Bangladesh to India . . . and then the Rest, *Himal South Asian*, October, http://www.himalmag.com/The-coming-crisis_nw3575.html, accessed 1 September 2010.

Chowdhury, R. 2009. Climate Change Induced Forced Migrants: In Need of Dignified Recognition Under a new Protocol, http://www.glogov.org/images/doc/equitybd.pdf, accessed 6 December 2013.

Christian Aid. 2007. Human tide: The real migration Crisis, http://www.christianaid.org.uk/Images/human-tide.pdf, accessed 3 January 2014.

Christoff, P. 1996. 'Ecological Modernisation, Ecological Modernities', *Environmental Politics*, 5 (3): 476–500.

Climate Action. 2014. Pollution Linked to 7 Million Deaths in 212 says WHO, http://www.climateactionprogramme.org/news/pollution_linked_to_7_million_deaths_in_2012_says_who, accessed 1 September 2014.

Climate Change, Nuclear Power and the Reframing of Risk in the UK News Media, *International Communication Gazette*, 73(107).

Clingerman, F. and O'Brien, K. J. 2014. 'Playing God: Why Religion Belongs in the Climate Engineering Debate', *Bulletin of the Atomic Scientists*, 70(3): 27–37.

CNA Corporation. 2007. National Security and the Threat of Climate Change, https://www.cna.org/sites/default/files/National%20Security%20and%20the%20Threat%20of%20Climate%20Change%20-%20Print.pdf, accessed 25 December 2014.

CNN News. 2010. U.N. Climate Chiefs Apologize for Glacier error, http://www.cnn.com/2010/WORLD/asiapcf/01/20/glacier.himalayas.ipcc.error/index.html, accessed 22 January 2014.
Collins, A. 2005. 'Securitization, Frankenstein's Monster and Malaysian Education', *The Pacific Review,* 18(4): 567–588.
Commoner, B. 1971. *The Closing Circle*, New York: Knopf.
Connell, J. 2003. 'Losing Ground? Tuvalu, the Greenhouse Effect and the Garbage Can', *Asia Pacific Viewpoint*, 44(2): 89–107.
Connolley, W. M. 2009. Fourier 1827: MEMOIR on the Temperature of the Earth and Planetary Spaces, http://www.wmconnolley.org.uk/sci/fourier_1827/fourier_1827.html, accessed 16 August 2012.
Connelly, M. and Kennedy, P. 1994. 'Must it be the Rest Against the West?' *Atlantic Monthly*: 61–91.
Cordner, L. G. 2014. 'Exploring Risks and Vulnerabilities: An Alternate Approach to Maritime Security Cooperation in the Indian Ocean Region', *Journal of Defence Studies*, 8(2): 21–43.
Crist, E. 2007. 'Beyond the Climate Crisis: A Critique of Climate Change Discourse', *Telos* 2007(141): 29–55.
Crutzen, P. J. 2006. 'Albedo Enhancement by Stratospheric Sulfur Injections: A Contribution to Resolve a Policy Dilemma?' *Climatic Change,* 77(3–4): 211–220.
CSE. 2011. 'Indian environment minister Jayanthi Natarajan gives hard hitting speech, receives standing ovation', http://www.cseindia.org/content/indian-environment-minister-jayanthi-natarajan-gives-hard-hitting-speech-receives-standing-o, accessed 24 January 2015.
CSE. 2013. COP 19, 'India Loses Momentum at Warsaw', http://cseindia.org/content/cop19-warsaw-india-loses-momentum-warsaw, accessed 30 November 2014.

D

D'Andrea, D. 2013. 'Global Warming as a Globalised Risk and Global Threat for Future Generations', in Daniel Innerarity and Javier Solana (eds) *Humanity at Risk: The Need for Global Governance*, New York: Bloomsbury: 107–120.
Dabelko, G. D. 2009. 'A Word of Caution on Climate Change and "Refugees"', in Geoffrey D. Dabelko (ed.) *Environmental Change and Security Program Report 13*, Woodrow Wilson International Center for Scholars.
Dabelko, G. D. and Dabelko, D. D. 1995. Environmental Security: Issues of Conflict and Redefinition, Woodrow Wilson Environmental Change and Security Project Report 1: 3–12.
Dalby, S. 1999. 'Threats from the South? Geopolitics, Equity and Environmental Security', in D. Deudney and R. Matthew (eds) *Security and Conflict in the New Environmental Politics,* Albany: State University of New York Press.
Dalby, S. 2000. Geopolitical Change and Contemporary Security Studies: Contextualizing the Human Security Agenda, http://kms1.isn.ethz.ch/serviceengine/Files/ISN/46517/...56E8.../WP30.pdf, accessed 18 August 2014.
Dalby, S. 2002. *Environmental Security*, Minneapolis: University of Minneapolis Press.
Dalby, S. 2003. 'Environmental Geopolitics: Nature, Culture, Urbanity', in Kay Anderson, Mona Domosh, Steve Pile and Nigel Thrift (eds) *Handbook of Cultural Geography*, London: Sage: 498–509.

Dalby, S. 2007. 'Antropocene Geopolitics: Globalisation, Empire, Environment and Critique', *Geography Compass*, 1(1): 103–118.
Dalby, S. 2009. *Security and Environmental Change,* Cambridge: Polity Press.
Dalby, S. 2010. 'Reconceptualising Violence, Power and Nature: The Next Twenty Years of Critical Geopolitics', *Political Geography*, 29(5): 280– 288.
Dalby, S. 2013. 'The Geopolitics of Climate Change', *Political Geography,* 37: 38–47.
Dasgupta, C. 2012. 'Present at the Creation: The Making of the UN Framework Convention on Climate Change', in N. K. Dubash (ed.) *Handbook of Climate Change and India: Development, Politics and Governance,* New Delhi: Oxford University Press.
David Newman. 2010. 'Territory, compartments and borders: Avoiding the Trap of the Territorial Trap', *Geopolitics* 15(4): 773–778.
Davis, M. 1999. *Ecology of Fear – Los Angeles and the Imagination of Disaster,* New York: Vintage Books.
Davissen, J. and Long, S. 2003. *The Impact of Climate Change on Small Island States,* Friends of the Earth, Melbourne: Australia.
De, S. 2005. *Illegal Migrations and the North-East: A Study of Migrants from Bangladesh,* New Delhi: Anamika Pub & Distributors.
Deleuze, G. and Guattari, F. 1987. *A Thousand Plateaus: Capitalism and Schizophrenia,* Minneapolis, MN: University of Minnesota Press.
Demeritt, D. 2001. 'The Construction of Global Warming and the Politics of Science', *Annals of the Association of American Geographers,* 91(2): 307–337.
Depledge, J. 2005. *The Organization of Global Negotiations: Constructing the Climate Change Regime,* Sterling, VA: Earthscan.
Derrida, J. 1977. 'Signature Event Context', Lecture delivered at a Montreal conference entitled "Communication," organized by the Congrès international des Sociétés de philosophie de langue francais.
DeSilva-Ranasinghe, S. 2013. 'Interview: Admiral Samuel J. Locklear, C-in-C, PACOM', *South Asia Defense and Strategic Review,* http://www.defstrat.com/exec/frmArticleDetails.aspx?DID=388, accessed 16 December 2014.
Dikeç, M. 2005. 'Space, Politics and the Political', *Environment and Planning D,* 23(2): 171–188.
Dobson, A. (ed.) 1999. *Fairness and Futurity: Essays on Environmental Sustainability and Social Justice,* Oxford: Oxford University Press.
Dodds, F., Higham, A. and Sherman, R. 2009. *Climate Change and Energy Insecurity: The Challenge for Peace, Security and Development,* London: UK: Earthscan Publications.
Dodds, K. 2012. 'Anticipating the Arctic and the Arctic Council: Preemption, Precaution and Preparedness', in T. S., Axworthy, T., Koivurova and W., Hasanat (eds) *The Arctic Council: Its Place in the Future of Arctic Governance,* Toronto: Munk-Gordon Arctic Security Program.
Dodds, K. J. 2000. *Geopolitics in a Changing World,* Harlow and New York: Prentice Hall.
Doherty, B. 2002. 'The Revolution in High Lane? Direct Action Community Politics in Manchester in the 1970s', *North West Labour History Journal,* 27: 60–64.
Doherty, B. and Doyle, T. 2006. 'Beyond Borders: Environmental Movements and Transnational Politics', *Environmental Politics,* 15(5): 697–712.
Doherty, B. and Doyle, T. 2011. *Transnational Solidarities: Friends of the Earth International.* Basingstoke: Palgrave Macmillan.
Doherty. B. and Doyle. T. 2014. *Environmentalism, Resistance and Solidarity: The Politics of Friends of the Earth International,* Basingstoke: Palgrave Macmillan.

Doyle, T. 1990. 'The Pot at the End of the Rainbow: The Political Myth and the Ecological Crisis', *Philosophy and Social Action*, 16(4): 47–61.

Doyle, T. 1998. 'Sustainable Development and Agenda 21: The Secular Bible of Global Free Markets and Pluralist Democracy', *Third World Quarterly*, 19(4): 771–786.

Doyle, T. 2001. *Green Power: The Environment Movement in Australia*, Sydney: University of New South Wales Press.

Doyle, T. 2002. *Environmental Movements*, London: Routledge.

Doyle, T. 2005. *Environmental Movements in Majority and Minority Worlds*, New Brunswick, NJ: Rutgers University Press.

Doyle, T. 2008. 'Crucible for Survival: Earth, Rain, Fire and Wind', in T. Doyle and M. Risley (eds) *Crucible for Survival: Environmental Security and Justice in the Indian Ocean Region*, Piscataway: NJ: Rutgers University Press.

Doyle, T. 2012. Educating for Climate Change in Friends of the Earth International, you tube video 21 November, https://www.youtube.com/watch?v=_0amwDddo-A, accessed 31 August 2014

Doyle, T., Alfonsi, A. and Robertson, P. 2014. 'Does an open-free market economy make in Australia more or less secure in a globalised world?, in Daniel Baldino, Andrew Carr and Anthony J. Langlois (eds) *Australian Foreign Policy: Controversies and Debates*, Melbourne: Oxford University Press.

Doyle, T. and Catney, P. 2012. (Series Editors) *Transforming Environmental Politics and Policy*, Farnham, Burlington: Ashgate.

Doyle, T. and Chaturvedi, S. 2010. 'Climate Territories: A Global Soul for Global South?' *Geopolitics*, 15(3): 516–535.

Doyle, T. and Chaturvedi, S. 2011. 'Securitising the Climate Refugee: Conceptualization, Categories and Contestations,' in J. S. Dryzek, R. B. Norgaard and D. Schlosberg (eds), *Oxford Handbook of Climate Change and Society*, Oxford: Oxford University Press.

Doyle, T., Chaturvedi, S. and Rumley, D. 2016. *The Rise and Return of the Pacific*, Oxford: Oxford University Press.

Doyle, T. and Doherty, B. 2006. 'Green Public Spheres and the Green Governance State: The Politics of Emancipation and Ecological Conditionality', *Environmental Politics*, 15(5): 881–92.

Doyle, T. and McEachern, D. 1997. *Environment and Politics*, Routledge: New York.

Drache, D. 2008. *Defiant Publics: The Unprecedented Reach of the Global Citizen*, Cambridge: Polity.

Dryzek, J., Dowens, D., Hunold, C., Scholsberg, D. and Hernes, H.K. 2003. *Green States and Social Movements: Environmentalism in the United States, United Kingdom, Germany and Norway*, Oxford: Oxford University Press.

Durand, J. and Massey, D. 1992. 'Mexican Migration to the United States: A Critical Review', *Latin American Research Review*, (3): 42.

Durban Group for Social Justice. 2004 http://www.carbontradewatch.org/durban/durbandec.html, accessed 20 November 2014.

Durieux, J. F. 2009. 'Climate Change and Forced Migration Hotspots', *Bonn Climate Talks*, UNHCR.

E

Eckersley, R. 1993. 'Free Market Environmentalism: Friend or Foe', *Environmental Politics*, 2(1): 1–19.

Eckersley, R. 2004. *The Green State: Rethinking Democracy and Sovereignty*, Cambridge, MA: MIT Press.
Edelman, M. J. 1978. *Space and Social Order*, Institute for Research on Poverty, University of Wisconsin.
Eggertsson, G. and Borgne, E. 2007. Dynamic Incentives and the Optimal Delegation of Political Power, IMF Working Paper, https://www.imf.org/external/pubs/ft/wp/2007/wp0791.pdf accessed on 1 September 2014.
Ehrlich, P. R. 1968. *The Population Bomb*, New York: Buccaneer Books.
Elden, S. 2009. *Terror and Territory: the Spatial Extent of Sovereignty*, Minneapolis: University of Minnesota Press.
Elliott, L. 2012. 'Climate Change and Migration in Southeast Asia : Responding to a New Human Security Challenge', Asia *Security Initiative Policy Series Working Paper* (20).
ENB. 1995. Summary of the first conference of the parties to the United Nations Framework Convention on Climate Change: 28 March-7 April, 12 (21). International Institute for Sustainable Development. Retrieved from http://www.iisd.ca/download/pdf/enb1221e.pdf, accessed 21 August 2014.
ENB. 1996. Summary of the second Conference of the Parties to the United Nations Framework Convention on Climate Change', *Earth Negotiation Bulletin*, 12(38): International Institute of Sustainable Development.
ENB. 1997. Summary of the Third Conference of the Parties to the United Nations Framework Convention on Climate Change, 12(76). International Institute for Sustainable Development.
ENB. 1998. Report of the fourth conference of the parties to the United Nations Framework Convention on Climate Change: 2–13 November 1998, 12(97). International Institute for Sustainable Development.
ENB. 2001. Summary of the resumed sixth session of the conference of the parties to the UN Framework Convention on Climate Change, 12(176). International Institute for Sustainable Development.
ENB. 2002. Summary of the eighth conference of the parties to the UN Framework Convention on Climate Change: 23 October–1 November, 12(209). International Institute for Sustainable Development.
ENB. 2007. Summary of the thirteenth conference of the parties to the UN Framework Convention on Climate Change and third meeting of the parties to the Kyoto Protocol: 3–15 December 2007, 12(354). International Institute for Sustainable Development.
ENB. 2009. Summary of the Copenhagen climate change conference: 12(459). International Institute for Sustainable Development, http://www.iisd.ca/climate/cop15/, accessed 23 August 2014.
ENB. 2011. Summary of the Durban climate change conference: 28 November – 11 December 2011, 12(534), International Institute for Sustainable Development.
ENB. 2012. 'Summary of the Doha Climate Change Conference: 26 November – 8 December 2012', 12(567), International Institute for Sustainable Development.
EPW. 2008. Editorial, *Economic and Political Weekly* (July 12, 2008).
EU, Australia and Norway also Sign up to new Carbon-Cutting Targets as Fortnight-Long Conference in Qatar Closes, The Guardian 8 December, http://www.theguardian.com/environment/2012/dec/08/doha-climate-change-deal-nations, accessed 1 September 2014.

F

Farbotko, C. and Lazrus, H. 2011. 'The First Climate Refugees? Contesting Global Narratives of Climate Change in Tuvalu', *Global Environmental Change,* 22(2): 382–390.

Faris, S. 2009. *Forecast: The Consequences of Climate Change, From the Amazon to the Arctic, from Darfur to Napa Valley,* New York: Henry Holt and Company.

Featherstone, D. 2013. 'The Contested Politics of Climate Change and the Crisis of Neo-liberalism', *Journal ACME,* 12.

Felli, R. 'An Alternative Socio-ecological Strategy? International Trade Unions' Engagement with Climate Change', *Review of International Political Economy,* iFirst, 2013 (to be published).

Ferguson, J. 1994. *The Anti-Politics Machine: Development, Depoliticisation, and Bureaucratic Power in Lesotho,* Minneapolis: University of Minnesota Press.

Ferguson, K. E. 2009. 'The sublime object of militarism', *New Political Science* 31(4): 475–486.

Ferris, E. G. 2008. 'Making Sense of Climate Change, Natural Disasters, and Displacement: A Work in Progress', *Refugee Watch,* (31): 75–90.

Financial Express. 2009. Climate Change Your Biggest Enemy: Dr. Pachauri Warns Indian Army, http://www.financialexpress.com/news/climate-change-your-biggest-enemy-dr.-pachauri-warns-indian-army/482641/, accessed 4 September 2013.

Flinders, M. and Buller, J. 2006. 'Depoliticisation: Principles, Tactics and Tools', *British Politics,* 1(3): 293–318.

Floyd, R. 2007. 'Human Security and the Copenhagen School's Securitization Approach: Conceptualizing Human Security as a Securitizing Move', *Human Security,* 5(winter): 38–59.

FoE Argentina. 2010. Translated from in Spanish, viewed November, 2010 http://www.amigos.org.ar/CambioClimatico/cambioclimatico2010.html.

FoE Australia. 2010a. Climate Justice: Our Climate Campaign in 2011, viewed October, 2010 http://www.foe.org.au/climate-justice.

FoE Australia. 2010b. Greens Call for Climate Refugees Welcome, viewed August 2010, http://www.foe.org.au/media-releases/2010-media-releases/greens-call-for-recognition-of-climate-refugees-welcomed.

FoE Belgium. 2010. Energy and Climate, Translated from French, viewed February 2010, http://www.amisdelaterre.be/spip.php?rubrique23.

FoE Chile. 2010. Translated from Spanish, viewed March 2010, http://www.webcodeff.cl/espanol/sitio/020.htm

FoE Cyprus. 2010. viewed July 2010, http://foecyprus.weebly.com/climate-justice.html.

FoE Europe. 2010. Demand Climate Justice: The Biggest Threat Our Planet is Facing!, viewed May 2010, http://www.foeeurope.org/climate/index.htm.

FoE EWNI. 2010a. Why Tackle Climate Change? viewed May 2010, http://www.foe.co.uk/campaigns/climate/climate_change_why.html.

FoE EWNI. 2010b. Unfair Burden of Climate Change, viewed July 2010, http://www.foe.co.uk/campaigns/climate/issues/climate_change_unfair.html.

FoE EWNI. 2010c. viewed May 2010, www.thebigask.eu.

FoE France. 2010. translated from French, viewed June 2010, http://www.amisdelaterre.org/-Position-des-Amis-de-la-Terre-sur,584-.html.

FoE Hungary. 2010. viewed July 2010, http://www.mtvsz.hu/programok_list_en.php?which=12

FoE Indonesia. 2010. viewed June 2010, http://walhi.or.id/en/campaign/climate-and-energy/176-siaran-pers/368-dont-trade-off-our-climate.

FoE International. 2002. Media Release 2002: Urgent Action needed to Address Climate Change, viewed November 2002, http://www.foe.org.au/media-releases/2002-media-releases/mr_28_10_02_2.htm.

FoE International. 2010. Chile: Building the Movement for Climate Justice, viewed April 2010, http://www.foei.org/en/resources/publications/annual-report/2008/what-we-achieved-in-2008/member-groups/latin-america-and-the-caribbean/chile-building-the-movement-for-climate-justice.

FoE Ireland. 2011. Climate Change, viewed February 2011, http://www.foe.ie/climatechange/>.

Foster, J. B. 2009. *The Ecological Revolution: Making Peace with the Planet*, New York: Monthly Review Press.

Foster, J. B. 2012. 'The Planetary Rift and the New Human Exemptionalism: A Political-Economic Critique of Ecological Modernization Theory', *Organization & Environment,* 25(3): 211–237.

Foster, J. B. and Clark, B. 2009. Ecological Imperialism: The Curse of Capitalism, http://www.nodo50.org/cubasigloXXI/taller/foster_clark_301104.pdf, accessed 3 September 2013.

Fox, W. 1990. *Towards a Transpersonal Ecology: Developing New Foundations for Environmentalism*, Albany: State University of New York Press.

Freedman, L. 1998. 'International Security: Changing Targets', *Foreign Policy,* (Spring): 48–64.

Freeland, C. 2012. *Plutocrats: The Rise of the Global Superrich and the Fall of Everyone Else*, Penguin Books.

French, D. and Scott, K. 2009. '"Implications of Climate Change for the Polar Regions": Too Much, Too Little, Too Late', *Melbourne Journal of International Law*, 10: 631–654.

Friends of the Earth. 2005. Citizen Guide to climate Refugee, http://www.foe.org.au/resources/publications/climate-justice/CitizensGuide.pdf, accessed 12 July 2014.

G

Gamble, A. 2000. *Politics and Fate*, Cambridge: Polity Press (in association with Blackwell Publishers).

Gamble, A. 2009. *The Spectre at the Feast: Capitalist Crisis and the Politics of Recession,* Basingstoke and New York: Palgrave Macmillan.

Gardiner, S. M. 2010. 'Is "Arming the Future" with Geoengineering Really the Lesser Evil? Some Doubts about the Ethics of Intentionally Manipulating the Climate System', in S.M. Gardiner, S. Caney, D. Jamieson. and H. Shue (eds) *Climate Ethics: Essential Readings*, Oxford: Oxford University Press: 284–314.

Gibbs, D. 2000. 'Ecological Modernisation, Regional Economic Development and Regional Development Agencies', *Geoforum* 31(1): 9–19.

Giddens, A. 2008. The Politics of Climate Change: National Responses to the Challenge of Global Warming, Policy Network Paper, http://www.fcampalans.cat/images/noticias/The_politics_of_climate_change_Anthony_Giddens(2).pdf, accessed 30 August 2014.

Giddens, A. 2009. *The Politics of Climate Change*, Cambridge: Polity Press.

Gilbert, E. 2012. 'The Militarization of Climate Change' *ACME: An International E-Journal for Critical Geographies'*, 11 (1): 1–14.

Gilboa, E. 2005. 'The CNN Effect: The Search for a Communication Theory of International Relations', *Political Communication*, 22: 27–44.

GOI. 2007. Developing World Cannot Accept a Freeze in Global Inequity: PM, http://www.pib.nic.in/newsite/erelease.aspx?relid=23764, accessed 30 August 2014.

GOI. 2008. PM's Speech on Release of Climate Change Action Plan, Prime Minister's Office (PMO), http://pmindia.nic.in/speeches.htm.

Goldman, R. K. 2009. 'The Guiding Principles: Normative Status and Its Effective Domestic Implementation', in Sibaji Pratim Basu (ed.) *The Fleeing People of South Asia: Selections from Refugee Watch*, London: Anthem Press.

Gong, Q. and Billon, P. L. 2014. 'Feeding (On) Geopolitical Anxieties: Asian Appetites, News Media Framing and the 2007–2008 Food Crisis', *Geopolitics* 19(2): 239–265.

Goodall, C. 2009. 'The green movement must learn to love nuclear power', *The Telegraph*, 23 February 2009, http://www.independent.co.uk/voices/commentators/chris-goodall-the-green-movement-must-learn-to-love-nuclear-power-1629354.html, accessed 12 November 2014.

Goods, C. 2011. 'Labour Unions, the environment and "green jobs"', *Journal of Australian Political Economy*, 67: 47–67.

Good, C. 2013. Back in the Climate Game: Obama Will Lay Out Global-Warming Plan, Tweets Video, 22 June, http://abcnews.go.com/blogs/politics/2013/06/back-in-the-climate-game-obama-will-lay-out-global-warming-plan-tweets-video/, accessed 31 August 2014.

Government of Bangladesh. 2009. Bangladesh Climate Change Strategy and Action Plan, Dhaka: Ministry of Environment and Forests, Government of the People's Republic of Bangladesh.

Grant, H., Randerson, J. and Vidal, J. 2009. UK Should Open Borders to Climate Refugees, Says Bangladeshi Minister, The Guardian, 4 December, http://www.guardian. co.uk/environment/2009/nov/30/rich-west-climate-change/print, accessed 8 September 2010.

Greenberg, J. 2000. 'Opinion Discourse and Canadian Newspapers: The Case of the Chinese Boat People', *Canadian Journal of communication*, 25(4): 517–537.

Gregory, D. 1994. *Geographical imaginations*, Oxford: Blackwell.

Gregory, D. 2004. *The Colonial Present: Afghanistan, Palestine, Iraq*, Oxford: Blackwell Publishing.

Gregory, D. 2009. 'Imaginative Geographies', in D. Gregory, R. Johnston, G. Pratt, M. J. Watts and S. Whatmore (eds) *The Dictionary of Human Geography*, Chichester: Wiley-Blackwell.

Greig, A., Hulme, D. and Turner, M. 2007. *Challenging Global Inequality: Development Theory and Practice in the 21st Century*, Basingstoke and New York: Palgrave Macmillan.

Grundmann, R. 2007. 'Climate Change and Knowledge Politics', *Environmental Politics*, 16 (3): 414–432.

Gupta, D. 2005. 'Migration Development and Security', in F. Dodds. and T. Pippard. (eds) *Human & Environmental Security: An Agenda for Change*, USA: Earthscan.

Gupta, P. 2008. 'From Chipko to Climate Change: Remote Rural Communities Grapple with Global Environmental Agenda', *Mountain Research and Development*, 28(1): 4–7.

Gupta, D. 2009. 'Climate of Fear: Environment, Migration and Security', in F. Dodds, A. Highham. and R. Sherman (eds) *Climate Change and Energy Insecurity*, London: Earthscan.

H

Haakon, L. 2009. 'Climate Change and Forced Migration Bangladesh', *IOP Conference Series Earth and Environmental Science*, 6(56): 562015.

Hajer, M. A. 1995. *The Politics of Environmental Discourse: Ecological Modernization and the Policy Process*, New York, NY: Oxford University Press.

Hajer, M. 1996. 'Ecological Modernisation as Cultural Politics', in S. Lash, B. Szersxynski. and B. Wynne (eds) *Risk, Environment, and Modernity*, London: Sage.

Hallding, K., Jürisoo, M., Carson, M. and Atteridge, A. 2013. 'Rising Powers: The Evolving Role of BASIC Countries', *Climate Policy*, 13(5): 608–631.

Hamilton C. 2013. *Earthmasters: The Dawn of the Age of Climate Engineering*, New Haven, CT: Yale University Press.

Hansen, J. 2005a. 'A Slippery Slope: How Much Global Warming Constitutes "dangerous anthropogenic interference"?: An Editorial Essay', *Climatic Change*, 68: 269–279.

Hansen, J. 2005b. Is there still time to avoid "dangerous anthropogenic interference" with global climate? A tribute to Charles David Keeling. Presentation given December, 2005, at the American Geophysical Union, San Francisco, accessed 1 October 2014, http://www.columbia.edu/!jeh1/

Hansen, J. 2010. *Storms of My Grandchildren: The Truth About the Coming Climate Catastrophe and Our Last Chance to Save Humanity*, London, Berlin, New York: Bloomsbury.

Harrison, G. 2005. 'The World Bank, Governance and Theories of Political Action in Africa', *The British Journal of Politics and International Relations*, 7(2): 240–260.

Harriss, J. 2002. *Depoliticizing Development: The World Bank and Social Capital*, London: Anthem Press.

Hartmann, B. 2010. 'Rethinking climate refugees and climate conflict: rhetoric, reality and the politics of policy discourse', *Journal of International Development* 22(2): 233–246.

Harvey, D. 2005. *A Brief History of Neo-Liberalism*, Oxford: Oxford University Press.

Harvey, S. 2006. 'International Borders: What They Are, What They Mean, and Why We Should Care', SAIS *Review*, 26(1): 3–10.

Harvey, F. 2012. Doha Climate Change Deal Clears Way for 'damage aid', to poor nations, 8 December, http://www.theguardian.com/environment/2012/dec/08/doha-climate-change-deal-nations, accessed 1 September 2014.

Hathaway, J. 2005. *The Rights of Refugees Under International Law*, New York: Cambridge University Press.

Hay, P. 2004. *Main Currents in Western Environmental Thought*, Sydney: University of New South Wales Press.

Hayek, F. A. 1944. The *Road to Safdom Fiftienth Anniversary,* Edition Chicago: The University of Chicago.
Hier, S. and Greenberg, J. 2002. 'Constructing a Discursive Crisis: Risk, Problematization and Illegal Chinese in Canada', *Ethnic and Racial Studies,* 25(3): 490–513.
Hindustan Times. 2006. Advani Lashes out at Congress over Illegal Immigration Issue, http://www.highbeam.com/doc/1P3-1017566601.html, accessed 26 January 2014.
Hinnawi, E. L. 1985. *Environmental Refugees,* Nairobi: United Nations Environment Programme.
Hobbes, T. 1651. *Leviathan,* Oxford: Clarendon Press.
Holger, S. 2007. 'Towards a Theory of Securitization: Copenhagen and Beyond', *European Journal of International Relations,* 13(3): 357–383.
Homer-Dixon, T.F. 1994. 'Environmental Scarcities and Violent Conflict: Evidence from Cases', *International Security,* 19(1): 5–40.
Hopkins, L. 2008. 'Mixing Climate Change With the War on Terror', *The Bulletin of the Atomic Scientists* 26 August, http://thebulletin.org/mixing-climatechange-war-terror-0, accessed 31 August 2014.
Hulme, M. 2008a. 'Geographical Work at the Boundaries of Climate Change', *Transactions of the Institute of British Geographers,* 33(1): 5–11.
Hulme, M. 2008b. 'The Conquering of Climate: Discourses of Fear and Their Dissolution', *The Geographical Journal,* 174(1): 5–16.
Hulme, M. 2009. *Why We Disagree About Climate Change: Understanding Controversy, Inaction and Opportunity,* Cambridge: Cambridge University Press.
Hunter, L. M. and David, E. 2009. 'Climate Change and Migration: Considering Gender Dimensions', in E. Piguet, P. de Guchteneire. and A. Pecoud (eds) *Climate Change and Migration,* New York, NY: UNESCO Publishing and Cambridge University.
Hyndman, J. 2012. 'The Geopolitics of Migration and Mobility', *Geopolitics,* 17(2): 243–255.
Hyvarinen, J. 2012. Loss and Damage Caused by Climate Change: Legal Strategies for Vulnerable Countries, Foundation for International Environmental Law and Development (FIELD), http://www.field.org.uk/papers/loss-and-damage-caused-climate-change-legal-strategies-vulnerable-countries, accessed 01 September 2014.
Hyvarinen, J. 2013. Advancing the Climate Change Negotiations: Issues to Consider. Foundation for International Environmental Law and Development (FIELD), Foundation for international Environmental law and development, http://www.africa-adapt.net/media/resources/841/field_advancing_climate_negotiations_jan_2013.pdf, accessed 23 August 2014.

I

Ibrahim, M. 2005. 'The Securitization of Migration: A Racial Discourse', *International Migration,* 43(5): 163–87.
IDMC. 2013. Global Estimates 2012. People Displaced by Disasters, Internal Displacement Monitoring Centre, Norwegian Refugee Council, http://reliefweb.

int/sites/reliefweb.int/files/resources/global-estimates-2012-may2013.pdf, accessed 31 August 2014.

IDMC. 2014. Global Overview 2014 People Internally Displaced by Conflict and Violence. Internal Displacement Centre, Norwegian Refugee Council, http://www.internal-displacement.org/assets/publications/2014/201405-global-overview-2014-en.pdf, accessed 31 August 2014.

Inglehart, R. 1990. *Culture Shift in Advanced Industrial Society*. Princeton: Princeton University Press.

Ingram, A. and Dodds, K. 2009. *Spaces of Security and Insecurity: Geographies of the War on Terror*, Surrey England: Ashgate Publishing Ltd.

Innerarity, D. 2013. 'Governing Global Risks', in Daniel Innerarity and Javier Solana (eds) *Humanity at Risk: The Need for Global Governance*, New York: Bloomsbury: 1–7.

International Organization for Migration. 2009. *Migration, Environment and Climate Change: Assessing the Evidence*, http://publications.iom.int/bookstore/free/migration_and_environment.pdf, accessed 1 September 2014.

International Union for Conservation of Nature and Natural Resources. 2000. *Vision for Water and Nature – Compilation of all Project Documents*, Canada: Island Press.

IPCC. 1990. First Assessment Report on Climate Change, http://www1.ipcc.ch/ipccreports/assessments-reports.htm, accessed 16 September 2013.

IPCC. 1995. Second Assessment Report on Climate Change: Synthesis of Scientific-Technical Information, http://www1.ipcc.ch/pdf/climate-changes-1995/2nd-assessment-synthesis.pdf, accessed 12 September 2013.

IPCC. 2007. Fourth Assessment Report on Climate Change: Working Group 1: Summary for Policymakers, http://www.ipcc.ch/pdf/assessment-report/ar4/wg1/ar4-wg1-spm.pdf, accessed 10 September 2013.

IPCC. 2010a. Statement on the Melting of Himalayan Glaciers, http://www.ipcc.ch/pdf/presentations/himalaya-statement-20january2010.pdf, accessed 2 March 2014.

IPCC. 2010b. Intergovernmental Panel on Climate Change Statement on the Melting of Himalayan Glaciers, Geneva, 20 January, http://www.ipcc.ch/pdf/presentations/himalaya-statement-20january2010.pdf, accessed 15 November 2010.

IPCC. 2012. Meeting Report of the Intergovernmental Panel on Climate Change Expert Meeting on Geoengineering [O. Edenhofer, R. Pichs-Madruga, Y. Sokona, C. Field, V. Barros, T.F. Stocker, Q. Dahe, J. Minx, K. Mach, G.-K. Plattner, S. Schlömer, G. Hansen, M. Mastrandrea (eds)]. IPCC Working Group III Technical Support Unit, Potsdam Institute for Climate Impact Research, Potsdam, Germany, 90.

IPCC. 2013. Climate Change 2013: The Physical Science Basis. Contribution of Working Group I to the Fifth Assessment Report of the Intergovernmental Panel on Climate Change [T. F. Stocker, D. Qin, G.-K. Plattner, M. Tignor, S. K. Allen, J. Boschung, A. Nauels, Y. Xia, V. Bex and P. M. Midgley (eds)]. Cambridge University Press, Cambridge, United Kingdom and New York, NY, USA.

IPCC. 2014a. Climate Change 2014: Impacts, Adaptation, and Vulnerability Technical Summary of Working Group II, http://ipcc-wg2.gov/AR5/images/uploads/WGIIAR5-TS_FGDall.pdf, accessed 24 August 2014.

IPCC. 2014b. Climate Change 2014: Mitigation of Climate Change. Contribution of Working Group III to the Fifth Assessment Report of the Intergovernmental Panel on Climate Change [O. Edenhofer, R. Pichs-Madruga, Y. Sokona, E. Farahani, S. Kadner, K. Seyboth, A. Adler, I. Baum, S. Brunner, P. Eickemeier, B. Kriemann, J. Savolainen, S. Schlömer, C. von Stechow, T. Zwickel and J. C. Minx (eds)]. Cambridge University Press, Cambridge, United Kingdom and New York, NY, USA.

ISDR. 2004. Living With Risk: A Global Review of Disaster Reduction Initiatives, International Strategy for Disaster Reduction, http://www.unisdr.org/we/inform/publications/657, accessed 15 March 2014.

Islam, R. 2009. Climate Change induced Disasters and Gender Dimensions: Perspective Bangladesh, http://www.monitor.upeace.org/archive.cfm?id_article=616, accessed 30 January 2014.

J

Jacobs, M. 1999. 'Sustainable Development as a Contested Concept', in A. Dobson (ed.) *Fairness and Futurity: Essays on Environmental Sustainability and Social Justice*, Oxford: Oxford University Press.

Jasparro, C. and Taylor, J. 2008. 'Climate Change and Regional Vulnerability to Transnational Security Threats in Southeast Asia', *Geopolitics*, 13(2): 232–256.

Jaworowski, Z. 2007. 'The Greatest Scientific Scandal of Our Time', *21st Century Science & Technology* (Spring/Summer): 14–28.

Jayasuriya, K. 2008. 'Regionalising the state: political topography of regulatory regionalism,' *Contemporary Politics*, 14(1): 21–35.

Jim, T. 2008. Sustainable Security: Developing a Security Strategy for a Long Haul, Centre for New American Security, http://www.cnas.org/files/documents/publications/Thomas_SustainableSecurity_April08.pdf, accessed 5 September 2014.

Jones, S. D. 2012. *Masters of the Universe: Hayek, Friedman, and the Birth of Neoliberal Politics by Daniel Stedman Jones*, Princeton NJ: Princeton University Press.

Jutila, M. 2006. 'Desecuritizing Minority Rights: Against Determinism', *Security Dialogue*, 37(2): 167–185.

K

Kakonen, J. 1994. *Green Security or Militarized Environment*, Aldershot and Brookfield: Dartmouth Publishing Company.

Kaldor, M. 2013. *Global Civil Society: An Answer to War*, Polity Press: Cambridge.

Kandlikar, M. and Sagar, A. 1999. 'Climate Change Research and Analysis in India: An Integrated Assessment of a South–North Divide', *Global Environmental Change*, 9 (2), 119–138.

Kaplan, R. 1994. 'The Coming Anarchy', *Atlantic Monthly*, 273(2): 44–76.

Kaplan, R. 2010. *Monsoon: The Indian Ocean and the Future of American Power*, New York: Random House.

Karim, Z., Hussain, Z. S. G., and Ahmed, A. U. 1999. 'Climate Change Vulnerability of Crop Agriculture', in S. Huq, Z. Karim, M. Asaduzzaman. and

F. Mahtab (eds) *Vulnerability and Adaptation to Climate Change for Bangladesh* Bangladesh: Kluwer Academic Publishers.

Kartha, S. 2011. 'Discourses of the Global South', in J. S. Dryzek, R. B. Norgaard. and D. Schlosberg (eds) *The Oxford Handbook of Climate Change and Society*, Oxford: Oxford University Press: 504–520.

Keane, J. 2001. Fear and Democracy, http://johnkeane.net/21/topics-of-interest/fear-and-democracy, accessed 31 August 2014

Kees Van Der Pijl. 2014. *The Discipline of Primacy: Modes of Foreign Relations and Political Economy*, London: Pluto Press.

Keith D. W. 2000. 'Geoengineering the Climate: History and Prospect', *Annual Review of Energy and the Environment,* 25: 245–284.

Kelly, P. M. 2000. 'Climate Co-Operation or Neo-Colonial Exploitation', *Watershed,* 7(1): 66–68.

Klein, N. 1999. *No Logo*, New York: Piacdor.

Klein, N. 2007. *The Shock Doctrine: The Rise of Disaster Capitalism*, New York: Metropolitan Books/Henry Holt.

Kotani, T. 2014. 'Building Bridges over the Sea: Shinzo Abe's visit is a good time to operationalise the India-Japan maritime partnership,' *Delhi*, digital edition, 14 January 2014.

Krasner, S. 1999. *Sovereignty: Organised Hypocrisy*, Princeton: Princeton University Press.

Krishnan, K. 1978. *Prophecy and Progress*, London: Allen Lane.

Kuhn, T.S. 1961. *The Structure of Scientific Revolutions*, Chicago: University of Chicago Press.

Kuhn, T. 1962. *The Structure of Scientific Revolutions*, Chicago: The University of Chicago Press.

Kuhn, T.S. 1970. *The Structure of Scientific Revolutions*, Chicago: Chicago University Press.

Kuhn. T. 1996. *The Structure of Scientific Revolutions, 3rd edition, Chicago: University of Chicago Press.*

Kundzewicz, Z. W., Mata, L. J., Arnell, N. W., Do¨ll, P., Kabat, P., Jime´nez, B., Miller, K. A., Oki, T., Sen, Z. and Shiklomanov, I. A. (2007). 'Freshwater Resources and Their Management', in M. L. Parry, O. F. Canziani, J. P. Palutikof, P. J. van der Linden and C. E. Hanson (eds) *Climate Change 2007: Impacts, Adaptation and Vulnerability. Contribu-tion of Working Group II to the Fourth Assessment Report of the Intergovernmental Panel on Climate Change*, Cambridge: Cambridge University Press: 173–210.

Kunzig, R. and Broecker, W. 2009. *Fixing Climate: The Story of Climate Science – and How to Stop Global Warming*, London: GreenProfile/Sort Of Books.

Kurtzman, J. 2009. The Low Carbon Diet: How the Markets Can Curb Climate Change, *Foreign Affairs,* http://www.foreignaffairs.com/articles/65237/joel-kurtzman/the-low-carbon-diet, accessed on 5 October 2014.

L

Lafferty, W. M. and Meadowcroft, J. 2000. *Implementing Sustainable Development: Strategies and Initiatives in High Consumption Societies*, Oxford: Oxford University Press.

Landsea, C. 2005. Chris Landsea Leaves in IPCC 2007, http://sciencepolicy.colorado.edu/prometheus/archives/science_policy_general/000318chris_landsea_leaves.html, accessed 23 July 2013.

Le. Monde, F. R. 2007. Le Réchauffement Climatique Pourrait Déclencher une Guerre Civile Mondiale, http://abonnes.lemonde.fr/cgibin/ACHATS/ARCHIVES/archives.cgi?ID=6fd940d00df097965f03510c435894afad5f7c82ad3698b8, accessed 13 October 2013.

Lee, P. 2010. 'Capitalism is Root Cause of Climate Change-President Evo Marales', http://www.foei.org/, accessed 21 April 2010.

Leighton, M. 2006. 'Desertification and Migration', in P. M. Johnson., K. Mayrand., M. Pacquin (eds) *Governing Global Desertification*, UK: Ashgate.

Leighton, M. 2010. 'Climate Change and Migration : Key Issues for Legal Protection of Migrants and Displaced Persons', Background Paper, Washington DC: German Marshall Fund of the United States.

Levy, A. M. 1995. 'Is the Environment a National Security Issue?' *International Security*, 20(2): 35–62.

Lilley, S. 2012. 'The Apocalyptic Politics of Collapse and Birth', in S. Lilley, D. McNally, E. Yuen, J. Devis (eds) *Catastrophism: The Apocalyptic Politics of Collapse and Rebirth*, Oakland: PM Press.

Lilley, S. McNally, D., Yuen, E. and Devis, J. 2012. *Catastrophism: The Apocalyptic Politics of Collapse and Rebirth*, PM Press: Oakland.

Lobo-Guerrero, L. 2011. *Insuring Security: Biopolitics, Security and Risk*, London: Routledge.

Locke, J. T. 2009. 'Climate Change-Induced Migration in the Pacific Region: Sudden Crisis and Long-Term Developments', *The Geographical Journal*, 175 (3): 171–180.

Loewenstein, A. 2009. Australia Snubs Tamil Refugees, http://www.thenation.com/doc/20091221/loewenstein, accessed on 23 December 2013.

Lohmann, L. 2005. 'Marketing and Making Carbon Dumps: Commodification, Calculation and Counterfactuals in Climate Change Mitigation', *Science as Culture* 14(3): 203–235.

Lohman, L. 2008. 'Carbon Trading, Climate Justice and the Production of Ignorance', *Development*, 51: 359–365.

Lomborg, B. 2009. *Cool it: The Skeptical Environmentalist Guide to Global Warming*, New York: Vintage Books.

Long, S. and Walker, C. 2005. Climate Refugees: The Hidden Cost of Climate Change, http://www.onlineopinion.com.au/view.asp?article=3569, accessed 23 December 2014.

Loster, M. L. and Warner, K. 2009. 'The Challenges of Climate and Migration', *Development and Cooperation*, 50: 323–325.

Lovbrand, E. and Stripple, J. 2006. 'The Climate as a Political Space: On the Territorialization of the Global Carbon Cycle', *Review of International Studies*, 32: 217–235.

Lövbrand, E. and Stripple, J. 2011. 'Making Climate Change Governable: Accounting for Sinks, Credits and Personal Budgets', *Critical Policy Studies*, 5(2): 187–200.

Lovelock, J. 2010. *The Vanishing Face of Gaia: A Final Warning*, London: Penguin Books.

Luciani, G. 1989. 'The Economic Content of Security', The *Journal of Public Policy*, 8(2): 151–173.

M

MacFarlane, S. N. and Khong, Y. F. 2006. *Human Security and the UN: A Critical History*, Bloomington: Indiana University Press.

Mahajan, S. 2007. Bangladesh: Strategy for Sustained Growth, Poverty and economic Management Network, *World Bank*, http://siteresources.worldbank.org/SOUTHASIAEXT/Resources/Publications/448813-1185396961095/4030558-1185396985915/summary.pdf, accessed 26 January 2014.

Marchart, O. 2007. *Post-Foundational Political Thought: Political Difference in Nancy, Lefort, Badiou and Laclau*, Edinburgh: Edinburgh University Press.

Marie, C. S. 2011. 'Chasing Climate Change–Exploring the Option of Assisted Migration', *The Forestry Chronicle,* 87(6): 707–711.

Martin, L. 1993. *Trade and Migration: NAFTA and Agriculture*, Washington, DC: Institute for International Economics.

Martin, S. 2010. 'Climate Change, Migration, and Governance', *Global Governance,* 16: 397–414.

Maslow, A. H. 1943. 'A Theory of Human Motivation', *Psychological Review*, 50: 370–396.

Maslow, A. M. 1954. *Motivation and Personality*. New York: Harper.

Mastanduno, M. 1998. 'Economics and Security Statecraft and Scholarship', *International Organization*, 52 (4): 825–54.

McAdam, J. 2011. 'Swimming Against the Tide: Why a Climate Change Displacement Treaty is Not the Answer', *International Journal of Refugee Law,* 23(1): 2–27.

McAdam, J. 2012. *Climate Change, Forced Migration, and International Law*, Oxford: Oxford University Press.

McCright, Aaron M. and Riley E. Dunlap. 2000. 'Challenging Global Warming as a Social Problem: An Analysis of the Conservative Movement's Counter-Claims', *Social Problems*, 47: 499–522.

McCright, A. M. and Dunlap, R. E. 2003. 'Defeating Kyoto: The Conservative Movement's Impact on US Climate Change Policy', *Social Problems,* 50(3): 348–373

McDonald, M. 2008.'Securitization and the Construction of Security', *European Journal of International Relations,* 14: 563–587.

McGoldrick, D., Williams, S. and Rajamani, L. 2010. 'The Making and Unmaking of the Copenhagen Accord', *International and Comparative Law Quarterly,* 59(03): 824–843.

McGregor, J. A. 1994. 'Climate Change and Involuntary Migration: Implications for Food Security', *Food Policy,* 19 (2): 120–132.

McKibben, B. 2006. *The End of Nature*, New York: Random House.

McLeman, R. 2008. 'Climate Change Migration, Refugee Protection and Adaptive Capacity-Building', *McGill International Journal of Sustainable Development Law and Policy,* 4(1): 1–18.

McMichael, A. J. 2009. Climate Change in Australia: Risks to Human Wellbeing and Health, http://www.globalcollab.org/Nautilus/australia/apsnet/reports/2009/australia-health.pdf, accessed 30 January 2014.

McMichael, C., Barnett J. and McMichael, A. J. 2012. 'An Ill Wind? Climate Change, Migration, and Health', *Environ Health Perspect*, 120: 646–654.

Mcsweeny, B. 1999. *Security, Identity and Interests, Cambridge*: Cambridge University Press.

Meadows, D.H. et al., 1972. *The Limits to Growth*. New York: Universe Books.

Meadowcroft, J. 2000. 'Sustainable Development: A New(ish) Idea for a New Century?' *Political Studies,* 48(2): 370–387.
Meadowcroft, J. 2002. 'Politics and Scale: Some Implications for Environmental Governance', *Landscape and Urban Planning* 61 (2–4): 169–79.
Meadowcroft, J. 2005a. 'Environmental Political Economy, Technological Transitions and the State', *New Political Economy,* 10(4): 479–498.
Meadowcroft, J. 2005b. 'From Welfare State to Ecostate', in J. Barry and R. Eckersley (eds) *The State and the Global Ecological Crisis,* MIT Press: 3–23.
Melissa, C. and Siu-lun, W. 2008. *Security and Migration in Asia: The Dynamics of Securitisation,* Oxon, USA: Routledge.
Mills, E. 2009. 'A Global Review of Insurance Industry Responses to Climate Change', *The Geneva Papers,* 34: 323–359.
Miller, M.A.L. 1995. *The Third World in Global Environmental Politics.* Buckingham: Open University Press.
Ministry of Environment and Forests Government of the People's Republic of Bangladesh. 2009. 'Climate Change Strategy and Action Plan (CCSAP)', http://www.moef.gov.bd/moef.pdf, accessed 15 January 2014.
Moberg, K. 2009. 'Extending Refugee Definitions to Cover Environmentally Displaced Persons Displaces Necessary Protection', *Iowa Law Review,* 94(3): 1107–1136.
MOEF. 2006. National Environmental Policy, http://www.moef.nic.in, accessed 3 July 2014.
Mohammed, A. 1984. 'Security in the Third World: The Worm about to Turn?', *International Affairs,* 60(1): 41–51.
Moïsi, D. 2009. *The Geopolitics of Emotion: How Cultures of Fear, Humiliation and Hope are Reshaping World,* London: The Bodley Head.
Mol, A. P. J. 2003. *Globalization and Environmental Reform: The Ecological Modernization of the Global Economy,* Cambridge MA: The MIT Press.
Mol, A. P. J. 1995. *The Refinement of Production: Ecological Modernization Theory and the Chemical Industry,* Utrecht: Van Arkel.
Mol, A. P. J. and Spaargaren, G. 2000. *Ecological Modernization Theory in Debate: A Review. Environmental Politics,* Great Britain: Frank Class Publisher.
Moomaw, W. 2009. The History of Climate Change and Climate Science, http://www.docstoc.com/docs/2186386/The-History-of-Climate-Change-and-Climate-Science, accessed 6 October 2013.
Moore, P. 2005. The Environmental Movement: Greens Have Lost Their Way, 13 June, http://greenspiritstrategies.com/the-environmental-movement-greens-have-lost-their-way/, assessed 22 August 2014.
Moore, P. 2005. Nuclear Statement to the US Congressional Committee, 28 April, http://www.ecosense.me/index.php/views-articles/8-views/174-issues-110, accessed 1 September 2014.
Morgan, M. G. and Ricke, K. 2010. *Cooling the Earth through Solar Radiation Management: The Need for Research and an Approach to Its Governance,* Geneva: International Risk Governance Council.
Mortreux, C. and Barnett, J. 2009. 'Climate Change, Migration and Adaptation in Funafuti, Tuvalu', *Global Environmental Change,* 19(1): 105–112.
Moses, W. J. 2006. *International Migration Globalization's Last Frontier,* London: Zed Books Ltd.
Moyo, D. 2012. *Winner Take All: China's Race for Resources and What It Means for the World,* New York: Basic Books.

Mumford, L. 1934. *Technics and Civilization*, New York: Brace and Company.

Mummery, J. 2012. 'Protecting the Global Common: Comparing three Ethico-Political Foundations for Response to Climate Change', *Borderlands e-journal*, 11(3): 1–31.

Myers, N. 1993. 'Environmental Refugees in a Globally Warmed World', *Bioscience*, 43(11): 752–761.

Myers, N. 2002. 'Environmental Refugees: A Growing Phenomenon of the 21st Century', *Philosophical Transactions of The Royal Society B*, 357(1420): 609–613.

Myers, N. 2005. Environmental Refugees: An Emergent Security Issue, http://www.osce.org/documents/eea/2005/05/14488_en.pdf, accessed 4 January 2014.

N

Narain, S. 2012. 'CoP18, Doha: An assessment A Gateway that leads nowhere', Centre for Science and Environment, http://www.cseindia.org/content/cop18-doha-assessment-a-gateway-leads-nowhere, accessed 20 October 2014.

National Society of Conservationists Hungary. 2010. *Green Ways to Develop Environment Friendly and Sustainable Projects*, file:///Users/sanjaychaturvedi/Downloads/GreenWays_fin.pdf, accessed 10 December 2013.

Neill, O. S. and Nicholson S. C. 2009.'Fear Won't Do It', *Science Communication*, 30(3): 355–379.

Nettle, C. 2014. *Community Gardening as Social Action*, Aldershot: Ashgate.

Newland, K. 2011. *Climate Change and Migration Dynamics*, Washington D.C: Migration Policy Institute.

Newman, D. 2000. 'Boundaries, Territory and Postmodernity: Towards Shared or Separate Spaces', in M. A. Pratt and J. A. Brown (eds) *Borderlands Under Stress*, U. A. E: Kluwer Dordrecht.

Newman, D. 2006a. 'Borders and Bordering Towards an Interdisciplinary Dialogue', *European Journal of Social Theory*, 9(2): 171–186.

Newman, D. 2006b. 'On Borders and Power: A Theoretical Framework', *A Journal of Borderlands Studies,*18(1).

Newman, E. 2009. Human Security, http://www.isacompss.com/info/samples/humansecuritysample.pdf, accessed 24 August 2014.

Newman, D. 2010. 'Territory, compartments and borders: Avoiding the Trap of the Territorial Trap', *Geopolitics* 15(4): 773–778.

Newman, D. (ed.) 2013. *Boundaries, Territory and Postmodernity*, London: Frank Cass Publisher.

Nicholls, R. J., Wong, P. P., Burkett, V. R., Codignotto, J. O., Hay, J. E., McLean, R. F., Ragoonaden, S. and Woodroffe, C. D. 2007. 'Coastal Systems and Low-Lying Areas. Climate Change, Impacts, Adaptation and Vulnerability', in M. L. Parry, O. F. Canziani, J. P. Palutikof, P. J. van der Linden and C. E. Hanson (eds) *Contribution of Working Group II to the Fourth Assessment Report of the Intergovernmental Panel on Climate Change*, Cambridge: Cambridge University Press.

Nieuwenhuis, M. 2013. 'Intervention–Cartographic Nationalism and Territorial Confusion in East Asia'. Oxford: Antipode Foundation, http://antipodefoundation.org/2013/05/14/intervention-cartographic-nationalism/, accessed 10 November 2014.

Nine, C. 2010. 'Ecological Refugees, States Borders, and the Lockean Proviso', *Journal of Applied Philosophy*, 27(4): 359–375.

Nizam. M [interview]. 2008, SAM, FoE Malaysia HQ, Penang, with Tim Doyle.
Nordhaus, W. 2007. The Stern Review on the Economics of Climate Change, http://nordhaus.econ.yale.edu/stern_050307.pdf, accessed 5 November 2013.
Nussmaumer, P. 2007. 'Working of Carbon Market', *Economic and Political Weekly* (28 July): 3081–3085.

O

O'Brien, K.L. and Leichenko, R.M. 2000. 'Double exposure: Assessing the impacts of climate change within the context of economic globalization', *Global Environmental Change*, 10(3): 221–232
O'Brien, K. L. and Leichenko, R. M. 2003. 'Winners and Losers in the Context of Global Change', *Annals of the Association of American Geographers*, 93(1): 89–103.
O'Brien, R. and Williams, M. 2010. *Global Political Economy: Evolution and Dynamics*, 3rd ed. Basingstoke and New York: Palgrave Macmillan.
Offe, C. 1985. 'New Social Movements: Challenging the Boundaries of Institutional Politics', *Social Research* 52(4): 817–868.
Okereke, C., Mann, P., Osbahr, H., Müller, B. and Ebeling, J. 2007. Assessment of Key Negotiating Issues at Nairobi Climate COP/MOP and What it Means for the Future of the Climate Regime, *Tyndall Centre Working Paper* No. 10 http://www.tyndall.ac.uk/sites/default/files/wp106.pdf, accessed 31 August 2014.
Oliver-Smith, A. 2012. 'Debating Environmental Migration: Society, Nature and Population Displacement in Climate Change', *Journal of International Development*, 24: 1058–1070.
O'Neill, S. and Nicholson-Cole, S. 2009. '"Fear Won't Do It" Promoting Positive Engagement With Climate Change Through Visual and Iconic Representations', *Science Communication*, 30(3): 355–379.
O'Rioordan, T. and Voisey, H. 1997. *Sustainable Development in Western Europe: Coming to Terms with Agenda 21,* London: Frank cass.
ÓTuathail, G. 1992. 'The Bush Administration and the "End" of the Cold War: A Critical Geopolitics of US Foreign Policy in 1989', *Geoforum,* 23(4): 437–452.
ÓTuathail, G. and Agnew, J. 1992. 'Geopolitics and Discourse: Practical Geopolitical Reasoning in American Foreign Policy', *Political Geography,* 11 (2): 190–204.
Overton, P. 2006. 'The Nuclear Solution', Channel Nine, http://sixtyminutes.ninemsn.com.au/stories/peteroverton/259389/the-nuclear-solution, accessed 15 December 2014.
Oxfam Briefing Paper. 2009. Suffering the Science Climate change, People, and Poverty, http://www.oxfam.org.uk/resources/policy/climate_change/downloads/bp130_suffering_science.pdf, accessed 25 July 2013.

P

Paddison, R. 2009. Some Reflections on the Limitations to Public Participation in the Post-Political City, *L'Espace Politique* [En ligne], 8 | 2009–2, mis en ligne le 30 septembre 2009, consulté le 06 octobre 2014. URL: http://espacepolitique.revues.org/1393; DOI: 10.4000/espacepolitique.1393, accessed on 6 October 2014.

Page, E. A. 2007a. *Climate Change, Justice and Future Generations*, Cheltenham: Edward Elgar.
Page, E. A. 2007b. 'Fairness on the Day after Tomorrow: Justice, Reciprocity and Global Climate Change', *Political Studies*, 55: 225–242.
Pain, R. 2009. 'Globalized Fear? Towards an Emotional Geopolitics', *Progress in Human Geography*, 33(4): 466–486.
Pain, R. 2010.'The New Geopolitics of Fear', *Geography Compass*, 4(3): 226–240.
Pain, R. and Smith, J. S. 2008. *Fear: Critical Geopolitics and Everyday Life*, Ashgate: Aldershot.
Painter, J. and Jeffrey, A. 2009. *Political Geography*, 2nd edition., Los Angeles: Sage.
Palazuelos, E. and García, C. 2008. 'China's Energy Transition: Features and Drivers', *Post-Communist Economies*, 20: 461–481.
Pan, J. 2012. 'The Future of the International Climate Regime from China's Perspective', in W. Wang. and G. Zheng (eds) *China's Climate Change Policies*, London and New York: Routledge: 1–17.
Parry, M. L., Canziani, O. F., Palutikof, J. P., van der Linden, P. and Hanson, C. E. (eds) in IPCC. (2007). *Summary for Policymakers. In: Climate Change 2007: Impacts, Adaptation and Vulnerability. Contribution of Working Group II to the Fourth Assessment Report of the Intergovernmental Panel on Climate Change*, Cambridge University Press: Cambridge.
Pellissery, S. 2011. Contestations on Climate Science in the Development Context: The Case of India, 101 *Washington Convention Center* https://aaas.confex.com/aaas/2011/webprogram/Paper3201.html, accessed 31 August 2014.
Perch-Nielsen, S. L., Sabine, Bättig, M. and Imboden, D. 2008. 'Exploring the Link Between Climate Change and Migration', *Climatic Change*, 91(3–4): 375–393.
Perlas, O. 2007. Politics and IPCC Global Warming Report, http://96.0.107.6/?q=node/view/2143, accessed 30 October 2013.
Peterson, T. C., Connolley, W. M. and Fleck, J. 2008.'The Myth of the 1970s Global Cooling Scientific Consensus', *Bull. Amer. Meteor*, (89): 1325–1337.
Pierson, P. 2001. 'Post-Industrial Pressures on the Mature Welfare States' in P. Pierson (ed.) *The New Politics of the Welfare State*. Oxford: Oxford University Press.
Piguet, E. 2008. 'Climate Change and Forced Migration: How can International Policy Respond to Climate-Induced Displacement?' UNHCR Evaluation and Policy Analysis Unit Research Paper 153, Geneva.
Piguet, E., Pécoud, A., and Guchteneire, P. D. (eds) 2011. *Migration and Climate Change*, Cambridge: Cambridge University Press.
Piguet, E., Pecoud, A. and Guchteneire, P. D. 2011. 'Migration and Climate Change: An Overview', *Refugee Survey Quarterly*, 30(3): 1–23.
Pijl, Kees van der. 2014. *The Discipline of Primacy: Modes of Foreign Relations and Political Economy*, London: Pluto Press.
Podesta, J and Ogden, P. 2007. 'The Security Implications of Climate Change', *The Washington Quarterly*, 31(1): 115–138.
PRC. 2007. China's National Climate Change Programme. National Development and Reform Commission People's Republic of China, June 2007, http://en.ndrc.gov.cn/newsrelease/200706/P020070604561191006823.pdf, accessed 1 August 2014.
PRC. 2009. China's Policies and Actions for Addressing Climate Change, 2009. Beijing: Information Office of the State Council of People's Republic of China,

http://www.ccchina.gov.cn/WebSite/CCChina/UpFile/File1324.pdf, accessed 1 August 2014.

Pred, A. 2007. 'Situated Ignorance and State Terrorism: Silences, W.M.D., Collective Amnesia, and the Manufacture of Fear', in D. Gregory and A. Pred (eds) *Violent Geographies: Fear, Terror and Political Violence*, New York: Routledge.

R

Raco, M., Henderson, S. and Bolwby, S. 2008, 'Changing Times – Changing Place: Urban Development and the Politics of Space-Time', *Environment and Planning A*, 40: 2652–2673.

Rahman, H. Z., Hossain, M. and Sen, B. (eds). 1996. *1987–94 Dynamics of Rural Poverty in Bangladesh*, Dhaka: Bangladesh Institute of Development Studies.

Rajamani, L. 2008. 'From Berlin to Bali and beyond: Killing Kyoto softly?' *The International and Comparative Law Quarterly*, 57(4): 909–939.

Rajamani, L. 2009. 'India and Climate Change: What India Wants, Needs, and Needs to Do', *India Review,* 8(3): 340–374.

Rajamani, L. 2012. 'The Changing Fortunes of Differential Treatment in the Evolution of International Environmental Law', *International Affairs*, 88(3): 605–623.

Raleigh, C. 2009. 'New Directions in the Climate Change-Conflict Literature', *Environment and Conflict in Africa Reflections on Darfur*, 63–72. University for Peace.

Raleigh, C., Jordan, L. and Salehyan, I. 2008. 'Assessing the Impact of Climate Change on Migration and Conflict', *World,* 24(3): 1–57.

Ramachandran, R. 2009. Climate Change and the Indian Stand, *The Hindu*, 28 July, http://www.thehindu.com/todays-paper/tp-opinion/climate-change-and-the-indian-stand/article240453.ece

Ranciere, J. 2007. *Hatred of Democracy*. London: Verso.

Renaud, F., Bogardi, J. J., Dun, O. and Warner, K. 2007. Control, Adapt or Flee: How to Face Environmental Migration?. Bonn: United Nations Institute for Environment and Human Security, http://www.unhcr.or.jp/TokyoRelease/2009/In_Search_of_Shelter_May2009.pdf, accessed 4 December 2013.

Reuveny, R. 2005. Environmental Change Migration and Conflict: Theoretical Analysis and Empirical Explorations, http://www.gechs.org/downloads/holmen/Reuveny.pdf, accessed 15 January 2014.

Reuveny, R. 2007. 'Climate Change-Induced Migration and Violent Conflict', *Political Geography,* 26(6): 656–673.

Revell, A. 2005. 'Ecological modernization in the UK: rhetoric or reality?', *European Environment*, 15(6): 344–361.

Reveron, S. 2009. Climate Change and National Security, http://www.acus.org/new_atlanticist/climate-change-and-national-security, accessed 20 December 2014.

Revkin, A. C. 2009. Hacked E-Mail Is New Fodder for Climate Dispute, http://www.nytimes.com/2009/11/21/science/earth/21climate.html?_r=0, accessed 31 August 2014.

Roberts, S., Secor, A. and Sparke, M. 2003. 'Neoliberal Geopolitics', *Antipode,* 35 (5): 886–897.

Robock, A. 2008. '20 Reasons why Geoengineering May be a Bad Idea', *Bulletin of the Atomic Scientists*, 64(2): 14–18.

Rodney, W. 1972. *How Europe underdeveloped Africa*, London: Bogle L Ouventure Publications.

Rogers, R. 1992. 'The Future of Refugee Flows and Policies', *International Migration Review*, 26(4): 1112–1143.

Rosanvallon, P. 2008. *Counter-Democracy: Politics in an Age of Distrust*, Cambridge: Cambridge University Press.

Rothe, D. 2011. 'Managing Climate Risks or Risking a Managerial Climate: State, Security and Governance in the International Climate Regime', *International Relations*, 25(3): 330–345.

Russell, P. and Hunter, P. 2010. 'Nuclear Resurgence Poised for Liftoff', *Engineering News-Record*, 264 (7): 11.

Ryan, B. 2013. 'Zones and Routes: Securing a Western Indian Ocean', *Journal of the Indian Ocean Region*, 9(2): 173–188.

S

Salehyan, I. 2005. 'Refugees, climate change and instability.' Presented at Human Security and Climate Change: International Workshop, Oslo, 21–23 June (conference).

Salih, M. 2009. *Climate Change and Sustainable Development: New Challenges for Poverty Reduction*. Cheltenham: Edward Elgar.

Salleh, A. 2011. 'Making the Choice Between Ecological Modernisation or Living Well', *Journal of Australian Political Economy*: Special Issue – Challenging Climate Change, 66: 118–143.

Samaddar, R. 1999. *The Marginal Nation: Transborder Migration from Bangladesh to West Bengal*, Dhaka: University Press.

Samaddar, R. 2009. 'Power, Fear, Ethics', in Sibaji Pratim Basu (ed.) *The Fleeing People of South Asia: Selections from Refugee Watch*, London: Anthem Press.

Sandel, M. 2012. *What Money Can't Buy: Moral Limits of Market*, London: Allen Lane.

Schario, T. and Pao, S. 2011. 'Pew Study: Department of Defense Accelerates Clean Energy Innovation to Save Lives, Money', Press Release 21 September 2011, Pew Project on National Security, Energy and Climate', http://www.pewenvironment.org/news-room/press-releases/pew-study-department-of-defense-accelerating-clean-energy-innovation-to-save-lives-money-85899364102#, accessed 26 November 2014.

Schreurs, M. A. 2012. 'Climate Change Politics in an Authoritarian State: The Ambivalent Case of China', in J. Dryzek and D. Schlosberg (eds) *Oxford Handbook of Climate Change and Society*, Oxford: Oxford University Press.

Schwartz, M. L. and Notini, J. 1994. Desertification and Migration: Mexico and the United States, Research paper, https://www.utexas.edu/lbj/uscir/respapers/dam-f94.pdf, accessed 23 August 2014.

Schwartz, P. and Randall, D. 2003. An Abrupt Climate Change Scenario and Its Implications for United States National Security, http://www.climate.org/PDF/clim_change_scenario.pdf accessed 12 August 2014. assessed on 1 September 2014.

Scrivener, D. 2002. 'Environmental Security', in J. Barry and E. G. Frankland (eds) *International Encyclopedia of Environmental Politics*, London: Routledge.

Sen, A. 2009. *Idea of Justice*, London: Penguin Books Limited.

Sengupta, S. 2012. 'International Climate Negotiations and India's Role' in N. K. Dubash (ed.) *Handbook of Climate Change and India: Development, Politics and Governance*, New Delhi: Oxford University Press.

Shapiro, J. 2001. *Mao's War Against Nature: Politics and the Environment in Revolutionary China*, Cambridge: Cambridge University Press.

Shapiro, J. 2012. *China's Environmental Challenges*, Cambridge: Polity Press.

Sharp, J. 2000. *Condensing the Cold War: Reader's Digest and American Identity*, Minneapolis: University of Minnesota Press.

Sheridan, G. 2011. "New Australia-US push deals India in to Pacific." *The Australian*. Accessed October 22, 2012. http://www.theaustralian.com.au/news/opinion/new-australia-us-push-deals-india-in-to-pacific/story-e6frg76f-1226139302534, accessed 10 November 2014.

Sheth, D. L. 2004. 'Globalization and the New Politics of Micro Movements', *Economic & Political Weekly*, 39(1): 56.

Siddiqui, T. 2009. Climate Change and Population Movement: The Bangladesh case, http://www3.ntu.edu.sg/rsis/nts/Events/climate_change/session4/Concept%20paper-Tasneem.pdf, accessed 4 January 2014.

Skodvin, T. and Andresen, S. 2009. 'An Agenda for Change in U.S. Climate Policies? Presidential Ambitions and Congressional Powers', *International Environmental Agreements: Politics, Law and Economics*, 9(3): 263–280.

Smith, J. 2007. 'Climate Change, Mass Migration and the Military Response', *Orbis*, 51(4): 617–633.

Smith, O. A. 2012. 'Debating Environmental Migration: Society, Nature and Population Displacement in Climate Change', *Journal of International Development*, 24: 1058–1070.

Smith, E. and Vivekananda, J. 2007. A Climate of Conflict: The Links Between Climate Change, Peace and War, London: International Alert, http://www.international-alert.org/pdf/A_Climate_Of_Conflict.pdf, accessed 12 December 2013.

Sommerville, M., Essex J. and Le Billon, P. 2014. 'The 'Global Food Crisis' and the Geopolitics of Food Security', *Geopolitics*, 19(2): 239–265.

Soroos, M. 1997. *The Endangered Atmosphere: Preserving a Global Commons*, Columbia: University of South Carolina Press.

Sparke, M. 2005. *In the Space of Theory: Postfoundational Geographies of the Nation State*, Minneapolis: University of Minnesota Press.

Sparke, M. 2007. 'Geopolitical Fears, Geoeconomic Hopes, and the Responsibilities of Geography', *Annals of the Association of American Geographers*, 97(2): 338–349.

Sparke, M. 2013. *Introducing Globalization: Ties, Tensions and Uneven Integration*, Oxford: Wiley Blackwell.

Spencer, H. 2006. 2 Million Displaced by Storms, *Washington Post*, 16 January, http://www.washingtonpost.com/wpdyn/content/article/2006/01/12/AR2006011201912.html, accessed 17 January 2014.

Sprout, H. and Sprout, M. 1957. 'Environmental Factors in the Study of International Politics', in A. D. Jackson William (ed.) *Politics and Geographic Relationships: Readings on the Nature of Political Geography*, Englewood Cliffs: Prentice-Hall.

Sprout, H. and Sprout, M. 1956. 'Man- Milieu Relationship Hypotheses in the Context of International Politics', Princeton University Center of International Studies.

State Environmental Protection Administration of China, 2006, 'Report on the State of the Environment in China', http://english.mep.gov.cn/down_load/Documents/200710/P020071023479580153243.pdf, accessed 20 December 2014.

Stern, N. 2006. *The Economy of Climate Change: Stern Review on the Economics of Climate Change* Cambridge: Cambridge University Press.

Streck, C. and Terhalle, M. 2013. 'The Changing Geopolitics of Climate Change', *Climate Policy*, 13(5): 533–537.

Stripple, J. 2008. 'Governing the Climate, (B)ordering the World', in L. Lundqvist and A. Biel (eds) *From Kyoto to the Town Hall: Making International and Climate Policy Work at the Local Level*, London: Earthscan: 137–154.

Swyngedouw, E. 2007. 'The Post-Politicalcity', in Bavo (ed.) *Urban Politics Now, Re-Imagening Democracy in Theneo-Liberal City*, Rotterdam: Netherlands Architecture Institute Publishers.

Swyngedouw, E. 2008. 'Civil Society, Governmentality and the Contradictions of Governance-Beyond-the-State', in J. Hillier, F. Moulaert and S. Vicari (eds) *Social Innovation and Territorial Development*, Ashgate: Aldershot.

Swyngedouw, E. 2009. 'The Antinomies of the Postpolitical City: In Search of a Democratic Politics of Environmental Production', *International Journal of Urban and Regional Research* 33(3): 601–620.

Swyngedouw, E. 2013. 'The Non-Political Politics of Climate Change', *Journal of ACME*, 12.

Szerszynski, B. and Galarraga, M. 2013. 'Geoengineering Knowledge: Interdisciplinarity and the Shaping of Climate Engineering Research', *Environment and Planning*, 45: 2817–2824.

Szerszynski, B., Kearnes, M., Macnaghten, P., Owen, R. and Stilgoe, J. 2013. 'Why Solar Radiation Management Geoengineering and Democracy Won't Mix', *Environment and Planning* A, 45: 2809–2816.

T

Taureck, R. 2006. Securitisation Theory – The Story so far: Theoretical Inheritance and What it Means to be a Post-Structural Realist, http://www.allacademic.com//meta/p_mla_apa_research_citation/1/0/0/1/7/pges100177/p100177-1.php, accessed 15 July 2013.

Teräväinen, T., Lahtonen, M. and Martiskainen, M. 2011. 'Climate Change, Energy Security and Risk–Debating Nuclear New Build in Finland, France and the UK', *Energy Policy*, 39(6): 3434–3442.

Terhalle, M. and Depledge, J. 2013. 'Great-Power Politics, Order Transition, and Climate Governance: Insights From International Relations Theory', *Climate Policy*, 13(5): 572–588.

TERI. 2009. Climate Change Your Biggest Enemy: Dr. Pachauri Warns Indian Army, New Delhi, press releases, http://www.teriin.org/index.php?option com_ pressrelease&task details&sid 156, accessed 6 September 2010.

The Council on Energy, Environment and Water (CEEW). 2014. http://ceew.in/newsDetails.php?id=258, accessed 30 June 2014.

The Environmental Justice Foundation. 2009. No Place like Home: Where Next for Climate Refugee, http://www.indiaenvironmentportal.org.in/content/no-place-home-where-next-climate-refugees, accessed on 23 December 2009.
Thomas, C. 1987. *In Search of Security: The Third World in International Relations*, Brighton: Harvester Wheatsheaf.
Thomas, J. 2008. 'Sustainable Security', Centre for New American Security, http://www.cnas.org/en/cms/?1924, accessed 10 October 2014.
Tilt, B. 2010. *The Struggle for Sustainability in Rural China: Environmental Values and Civil Society*, New York: Columbia University Press.
Tirman, J. 2004. *The Maze of Fear: Security and Migration After September 11th*, New York: New Press.
Toke, D. 2011a. *Ecological Modernization and Renewable Energy*, Basingstoke: Palgrave Macmillan.
Toke, D. 2011b. 'Ecological Modernization, Social Movements and Renewable Energy', *Environment Politics*, 20(1): 60–77.
Tol, R. S. J. 2007. 'Biased Policy Advice from the Intergovernmental Panel on Climate Change', *Energy & Environment*, 18(7, 8): 929–936.
Torgerson, D. 2006. 'Expanding the Green Public Sphere: Post-Colonialism Corrections', *Environmental Politics*, 15(5): 713–730.
Traufetter, F. 2007. The Age of the Climate Refugees? UN Global Warming Report, http://www.spiegel.de/international/world/0,1518,476062,00.html, accessed 21 November 2013.

U

Ullman, R. H. 1983. 'Redefining Security', *International Security*, 8(1): 129–153.
UNDP. 1994. *United Nations Development Programme: Human Development Report*, New York: Oxford University Press.
UNDP. 2004. *Reducing Disaster Risk: A Challenge for Development, Disaster Reduction Unit, Bureau for Crisis Prevention and Recovery*, New York: John S. Swift Co.
UNDP. 2007. *Human development Report: Fighting Climate Change: Human Solidarity in a Divided World* 2007/8, New York: Palgrave Macmillan.
UNHCR. 1999. Human Rights of Migrants Commission on Human Rights Resolution 1999/44, http://www.unhchr.ch/Huridocda/Huridoca.nsf/0/134ef623dad1ad1080256763005834fb?Opendocument, accessed 15 May 2010.
United Kingdom. 2013. The National Adaptation Program: Making the Country, https://www.gov.uk/government/uploads/system/uploads/attachment_data/file/20986 6/pb13942-nap-20130701.pdf, accessed 10 July 2014.
United Nations. 2009. Address by her Excellency Sheikh Hasina Hon'ble Prime Minister, Government of the People's Republic of Bangladesh, The United Nations, New York, 26 September, http://www.un.org/en/ga/64/generaldebate/pdf/BD_en.pdf, accessed on 15 November 2010.
United Nations. 1992. United Nations Framework Convention on Climate Change. United Nations. http://unfccc.int/resource/docs/convkp/conveng.pdf, accessed 5 August 2014.
US Commission on Immigration Reform. http://www.utexas.edu/lbj/uscir/respapers/dam-f94.pdf, accessed 3 October 2013.

US. Department of State. 2014. 'Remarks by John Kerry Secretary of State@america Jakarta', Indonesia: http://www.state.gov/secretary/remarks/2014/02/221704.htm, accessed 2 August 2014.

V

Valentine, J. 2005. 'Rancière and Contemporary Political Problems', *Paragraph,* 28(1): 46–60.

Vaughan, N. E. and Lenton, T. M. 2011. 'A Review of Climate Geoengineering Proposals', *Climatic Change,* 109(3–4): 745–790.

Veron, R and Majumdar, A. 2011. 'Micro-insurance through corporate-NGO partnerships in West Bengal; opportunities and constraints', *Development in Practice,* 21 (1): 31–41.

W

Waever, O. 1995. 'Securitization and Desecuritization', in R. D. Lipschutz (ed.) *On Security,* New York: Columbia University Press.

Wæver, O. 2000. *Security Agendas Old and New – And How to Survive them,* Buenos Aires: Paper presented for a workshop on, The Traditional and the New Security Agenda: Inferences for the Third World, Universidad Torcuato di Tella Buenos Aires.

Wald, M. 2006. 'Ex-Environmental Leaders Tout Nuclear Energy', New York Times, 26 August, http://www.nytimes.com/2006/04/25/us/25nuke.html, accessed 1 September 2014.

Walsham, M. 2010. 'Assessing the Evidence: Environment, Climate Change and Migration in Bangladesh. Change', *International Organization for Migration.*

Walt, S. 1991. 'The Renaissance of Security Studies', *International Studies Quarterly,* 35(2).

Waltz, K. 1979. *Theory of International Relations,* New York: Random House.

Wanliss, J. 2010. *Resisting the Dragon: Domination, Not Death,* Burke, VA: Cornwall Alliance for the Stewardship of Creation.

Ward, B. and Dubos, R. 1972. *Only One Earth: The Care and Maintenance of a Small Planet,* New York: W.W. Norton.

Ward, B. and Dubos, R. 1987 *Only One Earth: The Care and Maintenance of a Small Planet,* New York: W.W. Norton.

Warner, K. 2009. 'Migration: Adaptation to Climate Change or Failure to Adapt?', IOP Conf. Series: *Earth and Environmental Science,* 6: 562006.

Warner, K. et al. 2010. 'Climate Change and Migration : Reflections on Policy Needs', *MEA Bulletin* – Guest Article No. 64

Weart, S. 2003. The Discovery of Global Warming: The Carbon Dioxide Greenhouse Effect, http://www.aip.org/history/climate/co2.htm, accessed 4 October 2013.

WEDO. 2008. Gender, Climate Change and Human Security Lessons from Bangladesh, Ghana and Senegal, http://www.gdnonline.org/resources/WEDO_Gender_CC_Human_Security.pdf, accessed 15 January 2009.

Weiner, M. 1993. *International Migration and Security,* Boulder: Westview Press.

Weiner, M. 1995. *The Global Migration Crisis: Challenges to States and to Human Rights*, New York: Harper Collins.
Werrell, C. and Femia, F. 2014. The U.S. Military, 3D Printing, and a Climate Secure Future, Briefer no. 22, 27 January, The Centre for Climate and Security, http://climateandsecurity.files.wordpress.com/2012/04/the-u-s-military-3d-printing-and-a-climate-secure-future_briefer-22.pdf.
Werz, M. and Manlove, K. 2009. Climate Change on the Move, Climate Migration Will Affect the World's Security, Center for American Progress. http://www.americanprogress.org/issues/2009/12/on_the_move.html, accessed 8 December 2014.
White, G. 2011. *Climate Change and Migration: Security and Borders in a Warming World*, Oxford: Oxford University Press.
Wigley, T. M. L. 2006. 'A Combined Mitigation/Geoengineering Approach to Climate Stabilization', *Science*, 314(5798): 452–454.
Wihbey, J. 2008. Covering Climate Change As a National Security Issue, 17 July http://www.yaleclimateconnections.org/2008/07/covering-climate-change-as-a-national-security-issue/, accessed 11 August 2014.
Wikinews. 2006. World's Biggest Polluters Won't Cut Back on Fossil Fuel, http://en.wikinews.org/wiki/World's_biggest_polluters_won't_cut_back_on_fossil_fuel, accessed on 5 October 2014.
Williams, M. C. 2003. 'Words, Images, Enemies: Securitization and International Politics', *International Studies Quarterly*, 47(4): 511–531.
Williams, A. 2008. 'Turning the Tide: Recognizing Climate Change Refugees in International Law', *Law and Policy*, 30 (4): 502–529.
Williams, J. 2006. *The Ethics of International Borders: Drawing Shifting Lines in the Sand*, Basingstoke: Palgrave Macmillan.
World Bank. 2013. Turn Down the Heat: Climate Extremes, Regional Impacts, and the Case for Resilience, Potsdam Institute for Climate Impact Research and Climate Analytics, http://documents.worldbank.org/curated/en/2013/06/17862361/turn-down-heat-climate-extremes-regional-impacts-case-resilience-full-report, accessed 31 August 2014.
World Commission on Environment and Development, 1987. *Our Common Future*, Oxford: Oxford University Press.
World Metrological Organization. 1979. 'Scientific Advisory Board of World Climate Conference', http://unesdoc.unesco.org/images/0003/000376/037648eb.pdf, accessed 23 November 2009.
World Meteorological Organisation. 1986. Assessment of the Role of Carbon Dioxide and of Other Greenhouse Gases in Climate Variations and Associated Impacts, http://www.icsu-scope.org/downloadpubs/scope29/statement.html, accessed 4 October 2013.
World watch Institute. 2000. 'Melting of Earth's Ice Cover Reaches New High', http://www.worldwatch.org/node/1673, accessed 30 November 2013.

X Y Z

Yuen, E. 2012. 'The Politics of Failure Have Failed: The Environmental Movement and Catatrophism', in S. Lilley, D. McNally, E. Yuen, J. Devis (eds) *Catastrophism: The Apocalyptic Politics of Collapse and Rebirth*, Oakland: PM Press.

Žižek, S. 1992 *Looking Awry. Cambridge*, MA: MIT Press.

Žižek, S. 1999a. 'Carl Schmitt in the Age of Post-Politics', in C. Mouffe (ed.) *The Challenge of Carl Schmitt*, London: Verso.

Žižek, S. 1999b. *The Ticklish Subject: the Absent Centre of Political Ontology*, London: Verso.

Index

CPI Antony Rowe
Chippenham, UK
2016-12-27 19:30